The APBC Book of Companion Animal Behaviour

The APBC Book of Companion Animal Behaviour

Second Edition

Edited by David Appleby

SOUVENIR PRESS

Contents

List of Illustrations

11. Urine spraying – a natural method of communication for the cat, but unpleasant for humans if performed indoors

12. Some multi-cat households are made up of sibling pairs such as these, but in many cases there is no such basis to the feline community and cats are expected to live with total feline strangers

13. Rabbits are curious creatures and will play with and investigate objects in their environment

14. Garden and indoor toys can provide security, shade and stimulation

15. The rabbit's natural behaviour makes it possible to use a litter tray

16a and b. Creative thinking can provide environmental stimulation for dogs in a shelter environment

17. Environmental enrichment for the cat in a shelter should include different levels and places to hide

Introduction

Since I edited the first edition of this book the study of companion animal behaviour and the profession of pet behaviour counselling have continued to advance. Then, as now, there is a long way to go in our search for a full understanding of the various causes for the development of behaviour problems and the treatment methods that are most effective. This edition has been updated where appropriate to reflect the current thinking of those issues written by people who have to process this and other information during the course of treating behaviour problems.

As before 'Companion Animal Behaviour' focuses on the three main companion animals: the dog, cat and rabbit because these are the species that are most commonly presented for the treatment of behaviour problems. Horses are deserving of being included within these pages because there is much to discuss but the focus of this book is the companion animals with which we cohabit. To a large extent, however, the principles discussed in these chapters, such as ethology, learning theory, the importance of appropriate early experience, environment, emotional states and issues relating to welfare and pet ownership can be related to other species.

Although the dog has been overtaken by the cat in its popularity for being kept as a pet it is still the species most frequently seen in behaviour practices, this is reflected in the number of chapters that are dedicated to it. Furthermore, this edition comes at a time when there has been a partly media driven resurgence of dominance/leadership related approaches to canine behaviour problems combined with punishment/negative reinforcement based training. To address this issue a chapter related to more properly understanding our relationship with dogs has been added to this edition.

Although the measure of success in animal training and behaviour modification must include the development of a behaviour that we want or the cessation of one we don't want, it should also be measured by its impact on animal welfare and the pet-owner relationship. This book hopes to further the readers' understanding of the important issues.

John Fisher, who edited the previous APBC book, mentioned in his introduction that he felt honoured to edit such a book and that at times 'my editor's pencil has trembled at the responsibility'. In preparation for the first edition of this book I knew exactly how he felt and in that at least nothing has changed. I would like to thank my colleagues within the APBC for contributing or revising the chapters of this book.

David Appleby, *Editor*

1
Evolutionary Aspects of Canine Behaviour
Jan Hoole

Canine Ecology

Dogs are carnivorous, predatory and social animals, which from an ecological perspective are, like other canine species, near the top of the food web. Wolves, from whom dogs are descended, are very successful as predators, and as with all organisms, the ecological niche they occupy affects the morphology (physical appearance), and the behavioural and social structure of the animal. Since the domestic dog (*Canis familiaris*) is directly descended from the Grey Wolf (*Canis lupus*), its physiology and behaviour are strongly influenced by its canine ancestor. Although humans have radically altered dog morphology through selective breeding, the physiology and some aspects of instinctive behaviour remain relatively unaltered.

To make sense of the evolution of the dog it is helpful to consider the ecology (how the species interacts with its environment) and behaviour of the wolf. In addition it may be useful to briefly discuss the processes by which evolution takes place with particular reference to how natural selection can act on behaviour. We can then consider the original function of some aspects of behaviour in the wolf and how these have been translated into our domestic companions.

Wolf Ecology, Morphology and Social Systems

The most striking characteristic of the wolf is its adaptation to running. This is true of most of the dog family, much more so than any of the other large carnivores. Of all the Canidae, wolves have the longest legs and the greatest endurance when running down prey. They tend to hunt prey that is much larger than them, and so require help with the kill. Although it does happen from time to time, an individual wolf finds it difficult to bring down an adult reindeer or musk ox, and impossible to tackle moose or bison. By hunting co-operatively they can isolate a weak animal from the herd and join together to bring it down and kill it.

This is not to say that wolves can be considered as small animals, since they are between 26 and 32 inches tall at the shoulder, and males can be as much as six feet five inches long from nose to tail tip. There is however great variation, not only between species of wolves, but also within species, in size, shape, coat and behaviour. Since wolves are exceedingly difficult to observe in the wild little is known about the behaviour of many species, but the Grey Wolf has been the subject of many studies and its social and predatory behaviour has been reasonably well documented. Much of the work has been done on captive and semi-wild populations. However, because of their artificial nature these studies are of limited value (see chapter 8) and they can only be interpreted as behaviour in abnormal social and environmental contexts.

Status (Uninhibited/Inhibited Social Behaviour)

The social structure of a wolf pack tends to remain reasonably stable over quite long periods. The structure is hierarchical and is based on individual ability to gain and retain access to resources. The male and female wolves with the greatest ability tend to form a pair and to lead the pack. The common view is that they are a breeding pair, and are the parents of all the other pack members. However, in artificially constructed semi captive packs there are several recorded instances, in which another male (usually called by the researchers the second ranked or beta male) was actually the sire of all the pups produced, while the "alpha" male did not successfully mate at all. In at least one of these situations, when the alpha male was removed and the beta male became the leader, he ceased to mate with the alpha female and the new beta male became the breeding male. It is not known if this situation happens in the wild.

The social structure within packs appears to have subcultures based on male and female lines. It appears that within these lines there is a fairly simple resource based hierarchy which becomes apparent as individuals reach social maturity. Pups and immature animals have an additional social structure, which is not determined by gender. Some adults may not even be part of the social structure of the pack and remain on the outskirts. Whether dogs live in social group of the same species, as may hounds, huskies and feral dogs or live in a mixed 'pack' with humans the social structure of the group they live in is established and maintained by communication.

Communication

Being large and well armed animals, if wolves fought over resources within the pack they could do each other serious injuries, and this would not contribute to the overall health of the pack as a reproductive unit. Encounters within the pack are often governed by ritualised signals and symbolic postures that help the pack members to interpret each other's status and state of mind and thus avoid genuine aggression. In order to maintain harmony and stability within the pack, wolves are very good at reading each other's body language and posture and they appear to use their own bodies to signal their state and intentions to other wolves. Whether or not such signalling is conscious or intentional is a matter for much discussion among academics. However, even if the signals are simply an unconscious response to the animal's internal state, they may enable other individuals to assess the situation and make a judgement about the likely behaviour of the animal.

Dogs maintain an excellent ability to read the signals emitted by humans and other dogs, but selective breeding has altered many of the structures used for signalling. Ear shape is frequently changed from pricked to long and floppy, with which it is difficult to signal inhibited behaviour. Paedomorphosis, (having the appearance of a new-born pup), is a popular characteristic in modern breeds, such as Cavalier King Charles spaniel, Schi Tzu, boxer and Chihuahua, and causes the dog to have a short muzzle, very round, prominent eyes, and often, dropped ears. Some breeds have very curled tails, or have their tails docked at birth, compounding the difficulties of signalling to other dogs.

A study carried out in 1997 found that the degree of paedomorphosis in individual breeds was an indication of the wolf developmental stage from which the signalling repertoire displayed by the breed was taken (1). Those breeds that most closely resemble adult wolves have the greatest number of wolf signals, some of which are attained by wolves in the later stages of development, while breeds that look more like young pups have fewer and more infantile signals.

It is also important to remember that many centuries of domestication separate dogs and wolves, and although they are genetically sufficiently similar to interbreed and produce fertile offspring, they are in fact, different species.

What is evolution?

The process of evolution operates on all living organisms over long periods of time, usually many thousands or millions of years. In the wild the process is thought to be mediated by natural selection, but in the case of domestic animals the selection is artificial, the pressure applied by human intention rather than by natural forces such as changes in the environment. By this route all the domesticated animals known to humankind have emerged, often from an ancestral species that is now extinct. In the case of the dog, its ancestor the grey wolf is still around, hanging on by a claw in the wilder areas of the earth.

Natural selection is a phrase that is often used to describe the means by which species arise. As a process it operates on all aspects of an organism, physical, ecological and behavioural. In order to understand how certain behavioural traits of the dog arose, it is necessary to consider the nature of its ancestor the wolf, and the possible evolutionary route by which the wolf acquired certain behaviours still observable in the dog. Evolution whether by natural or artificial selection will be more easily explained in the light of a basic understanding of how traits are inherited. Some possible routes for domestication of the dog are also worth considering.

Inheritance

The basic unit of inheritance is the gene. Every cell that has a nucleus in a living organism contains all the genetic information needed to replicate the whole organism.

Sometimes, in the process of copying or translation the genetic information or of building the cells, an error occurs. Fortunately the cell has efficient mechanisms for spotting and repairing such mistakes, and most of them are repaired straight away. Occasionally an error is missed, and this is called a mutation. The effect on the organism will vary depending on where on the gene such a mutation is situated. In some areas of the DNA a mutation may not affect the organism at all. However, if the mutation happens to be in the sperm or egg by which an organism reproduces, the effect will be felt by the next generation rather than by the existing organism.

A mutation in a germ cell may result in a fertilised egg that is not viable, or cause the new plant or juvenile animal to have a genetic defect which will mean that it is not able to survive to reproductive age or is sterile. In these cases the defect will not be passed on. The mutation may, however, be advantageous and give the organism an advantage in survival or reproductive strategy over other individuals in the population. These are the mutations that tend to be inherited and to become part of the gene pool of an organism.

The passage of genetic information is essentially a one way process. The sequence of the DNA in the nucleus of a cell can affect the size and shape of the molecules produced, but the size and shape of the molecule does not influence the sequence of the DNA stored in the nucleus. It is not possible for a learned behaviour to be inherited, because the changes brought about in the nervous system by learning cannot be fed back into the genes. However, there is the possibility that an organism may be genetically predisposed to have the ability to learn a particular behaviour or type of behaviour, and this can be inherited. If the environment in which the animal finds itself offers the opportunity to learn a behaviour that ultimately contributes to the animal's reproductive success, the animal is more likely to be able to learn it. This will allow it to have more offspring, some of which will carry the gene for being able to learn to the same thing.

It is a general genetic predisposition to behave in a certain way that is thus passed on and over generations is selected for in such a way that it becomes more narrowly adapted to the environmental conditions or ecological niche of an organism. Eventually the animal may become so behaviourally and morphologically adapted to its niche that it is incapable of behaving in a different way or of surviving in a different environment. It may also lose the ability to mate with organisms that are not so adapted and a new species is created.

Selection Pressure and Variation

It is clear therefore that natural selection can only operate if there is variation within a species. If all the organisms within a species are identical, as in clones, all will respond in the same way to changes in their environment and none will have an advantage. Some of the small variations between individuals may not be apparent when the environment is favourable, but may cause an individual to behave in a slightly different way under differing conditions.

Arctic Wolves generally sleep a short distance apart from each other at night, tightly curled into a ball with their noses covered by their tails. It might be expected that they would huddle together for warmth in the bitter arctic night, but David Mech points out that if they did this their breath would cause each other's coats to become damp and to freeze to the ground. It is easy to imagine that individuals that preferred to sleep close together would more often die from this cause before reaching maturity, therefore failing to pass on their genes. Those that had a tendency to be wary of their fellows and preferred to sleep a little way apart would survive and pass on the inclination for aloof behaviour to their offspring.

Kin Selection

In a wolf pack, for most of the time only one pair produces pups. Occasionally a lower ranking female will mate and succeed in bearing her pups, but generally a pack consists of two or three older animals and their offspring from successive years. The individuals that have not mated nevertheless take a share in the care of the current litter, often bringing food for them, helping to guard them and even to move them when the dam decides to change dens. This altruistic behaviour may seem difficult to understand. Surely they should be putting their efforts into attempting to pass on their own genes.

In fact, a mammal carries an average of 50% of the

same genes as its parents or full siblings and 25% of an aunt, uncle or grandparent. Another way of putting this is to say that the probability of an individual sharing a particular gene with its parents, offspring or siblings is 0.5, and with its aunts, uncles and grandparents is 0.25. This measure is called the coefficient of relatedness. Under some circumstances it is more advantageous to an individual's fitness, in terms of its chances of passing on its genes, for the animal to help its close relatives to survive and reproduce than to try itself to reproduce when there is small chance of success. This altruistic behaviour of helping relatives to survive or reproduce is called kin selection, and is the most likely evolutionary reason why wolves live in packs.

A lone pair of wolves has less chance of rearing pups to maturity in a harsh environment because it is more difficult for them to hunt large, energy-rich prey. While the female is in whelp and is nursing the pups the male must hunt alone and, although a single wolf is capable of killing an adult deer, caribou or even an elk, the risk involved in such encounters is greater for a lone wolf than for a pack. If the male is killed or injured the female would eventually have to leave the pups in order to hunt, leaving them vulnerable to predators. However, if the male only hunts small prey, such as arctic hare or beaver, the risks are reduced but the energy expended in hunting per prey item is likely to be much greater. By remaining with the parents and helping them to raise other pups, an individual wolf increases the chance of some of its genes surviving in the population.

One of the prerequisites for kin selection is that an animal is able to recognise its relatives, and there is evidence that some species, such as mice, are able to do this by recognising differences in a protein that occurs on all cells, the major histo-compatibility complex (MHC). It is believed that animals are able to recognise slight differences in scent which may correspond to differences in the MHC. It is possible that they can also detect similarities within a kinship line, and thus to whom they are related.

Wolves may also use visual cues to help recognise each other. Anyone who has kept more than one individual of the same breed of dog will know that, although they may look the same to outsiders, to the eye of familiarity it is scarcely possible to mix them up. Indeed it is possible to distinguish individuals by small details such as the pitch of a bark or the sound of a particular dog taking a drink.

Domestication

One of the more intriguing questions about the wolf concerns how this large and potentially dangerous predator became engaged with humans to such an extent that it eventually turned from *Canis lupus*, the wolf, into *Canis familiaris*, the domestic dog. In addition to considerable argument as to how this happened, biologists have expended a great deal of energy discussing when the event occurred.

The earliest identifiable dog remains in the fossil record, which were found in Northern Europe, are only 31,700 years old, while the oldest complete skeleton that can be said to be that of a dog was found in the grave of a woman in Israel, dated at 12,000 years ago. The use of molecular technology has enabled scientists to examine many theories about the origins of the dog, and a surprisingly large range of possibilities has been the result. Some studies have broadly agreed with the fossil record, putting the date of the earliest domestication event at between 12,000, and 16,300, years ago. Less conservative estimates suggest that it may have been as long as 100,000 years ago, though these tend to be the results of some of the earlier DNA studies.

The consensus as to the genetic origins of the dog is that the grey wolf *Canis lupus* is the common ancestor of all domestic dogs. Even the most ancient breeds found in the New World, such as the Xoloitzcuintli or Mexican hairless dog, came from stock imported from Eurasia. The most recent studies suggest that domestication occurred first in the 'fertile crescent' of the Middle East more than 12,000 years ago(2). Previous research using mitochondrial DNA, which is passed down only through the maternal line, had suggested South China as a point of origin and a time of no more than 16,300 years ago(3).

Perhaps it is of even more interest to speculate on the 'how' of domestication. It is easy to imagine that ancient hunters, having killed an adult wolf, could have taken the pups home to rear by hand and thus provided the first nurturing interaction between man and wolf, but less easy to imagine why they might do such a thing. While we may view wolves with sympathy in the present time, to a hunter in ancient times they would simply have been dangerous rivals, subsisting on similar prey to that utilised by humans. Unless one had already had experience of the process of socialisation of a wolf pup to humans and the profound effects it can have on the relationship between the two, to guess that socialisation is possible would be a leap of the imagination that would be beyond most humans. There are far more pressing reasons not to introduce a large predator into the family home, however appealing it may be at the puppy stage.

However, the effects of socialisation on wolf pups have been well-documented, and it has been shown that hand-rearing can produce wolves that are not only tame but very friendly towards people. Once such wolves were established in a human settlement and began to breed, the pups would naturally be in contact with people from birth and so would come to view them as members of the pack. Over generations, the wolves that would survive would be the ones that remained tame, co-operated with the humans and had a high degree of inhibited behaviour. Individuals that showed aggression would be killed. This sort of selection process can profoundly alter unrelated parts of the genome. Over millennia, the dog has become an animal that comes into oestrus (becomes receptive to mating) twice as often as a wolf, has a smaller brain to body mass ratio, less powerful jaws, compacted teeth, and is more juvenile in its behaviour than its lupine ancestor.

An alternative theory is that when humans began to live in camps and settlements, opportunist wolves may have scavenged the refuse that would accumulate in the area. These would naturally be animals that were sufficiently bold to come into the proximity of a camp, and in a hard season it is possible that the extra food source would help them to breed successfully, possibly with other scavenging wolves. The offspring of such wolves would undoubtedly learn from their parents about the supplementary food to be found in the vicinity of humans, and habituation from an early age would probably encourage them to go ever closer to the camps. Over a number of generations such wolves would eventually begin to show stronger traits of boldness and tameness, until it became possible for humans to approach them and form relationships with them.

This theory may be borne out by a long-running experiment carried out in Russia on silver foxes. These were bred for many generations, the only criterion for allowing them to breed being tameness. Another group of foxes were bred for aggression. After many generations individuals appeared that were dog-like in temperament. They seemed to actively seek human company and were quite playful. These traits bred true and produced a line of foxes that resembled dogs in their tameness. It seems that this was due to a mutation that arose in one particular fox which produced a sudden change in levels of tameness, rather than to a gradual increase in tameness over generations. It is interesting that, although appearance was never bred for in the foxes, several dog-like characteristics, such as a white tip to the tail, seemed to be present in the line of tame individuals that had not been seen in the original stock. Clearly not one gene but several areas of the genome have been affected through breeding for one character, and this illustrates how closely linked genes can be that code for entirely different aspects of the organism.

Another circumstance that seems to support this theory has been documented in a city in Transylvania. A female wolf moved into woodland overlooking the city and began to raise a litter. She was radio-collared by some scientists who wished to monitor her feeding habits. They found that she not only visited the local shepherds with their flocks every night in the hope of finding a stray sheep, but had also taken to making regular trips to the municipal dump to scavenge refuse. For a lone wolf, this is an easier option than attempting to catch large prey that could cause injury, or small prey that is very energy intensive to catch and gives low returns.

Behavioural similarities between dog and wolf

By whatever route domestication occurred, it is undeniable that many of the behavioural traits that we see in the dog are a modified version of what can be observed in the wolf. One of these is the greeting ritual. Most dogs, when their owners return home, even after a short absence, greet them with great enthusiasm, wagging the tail, wriggling and often attempting to jump up to lick the owner's face. In wolves a similar ritual is seen when the leader rejoins the pack after an absence. The other wolves will gather round the high status individual, licking and nipping his or her muzzle, wagging their tails and displaying unchallenging body language. An individual wolf will also greet a more assertive individual in this manner, and the pack sometimes engages in a group 'greeting ritual' before a hunt.

The origins of this behaviour seem to lie in the food begging that pups engage in between weaning and the age when they are old enough to feed from the carcass. For this period adult wolves bring food to the pups, carrying it in the stomach. The pups stimulate the adults to regurgitate the food for them by frenzied licking of the adult's mouth. It seems that this behaviour is carried on into adult-hood as a ritualised greeting.

When a dog wishes to engage in a greeting ritual with the human members of its 'pack' it has the major disadvantage that its face is between three and six feet lower than that of its owner, depending on breed of dog and height of human. The dog naturally tries to get as close as possible to the mouth of the person and jumps up to reach it. After initial reinforcement during puppy-hood, this is not usually well received by the humans, and can become a problem behaviour.

Vocal communication in both dogs and wolves is complex. Although vocal communication is often described as barking, growling, whining and howling, this does not adequately indicate the range of sounds produced and the information apparently conveyed by them. One of the most recognisable sounds produced by wolves is the howl. In Canada at certain seasons hundreds of people drive many miles into the wilderness to hear wolves howling, and recordings of 'wolf song' have been popular for many years in the music industry.

Howling in wolves occurs in several different circumstances. When the pack is separated an individual will often howl and this attracts other members to it. Packs may also howl before a hunt, and this seems to increase the social cohesion of the pack at a time when the need for co-operation may be at odds with the individual's instinct to hunt for itself. In captive wolves it has been observed that the amount of howling is correlated with the degree of isolation of the wolf from people and from other wolves. It has been suggested that howling in these circumstances is an expression of loneliness.

It is no co-incidence that the most frequent reports of howling in domestic dogs occur when the dog is left alone for long periods. Separation problems are dealt with elsewhere, but it is worth noting that this behaviour is directly traceable to the behaviour of re-uniting scattered pack members after hunting in wild wolves. Although we may suppress the behaviour in our companion animals by training them to cope with being alone, the unease that is generated in social animals by isolation is still present in dogs.

Herding behaviour commonly displayed by the breeds such as the Border Collie can also be traced to its origins in the hunting behaviour of the wolf. The hunting sequence in a predator can be separated into several distinct parts. It begins with the animal displaying a fixed gaze towards the prey, known as 'eye', moving on to stalk, chase, pounce and kill. In some breeds of dog individual components of the sequence have been exaggerated by selective breeding until they are displayed to the exclusion of other parts. In the collie, 'eye' and 'stalk' are the most intensely displayed components, with the result that sheep faced with an apparent predator tend to bunch together and move away. The collie will often run round a herd of sheep or flock of geese, and this behaviour may have its origins in the predatory intention to find the weak animal in the herd. Certainly, the greatly prized ability to 'cut out' marked sheep in the shedding ring at sheepdog trials can be seen as a reflection of the wolf behaviour of singling out the vulnerable prey animal. However, although working sheepdogs are trained to, and may be genetically predisposed to, inhibit the latter parts of the predatory sequence, especially 'pounce' and 'kill', it should not be forgotten that it remains part of the behavioural repertoire of the breed, as of many other breeds.

Even if the dog may have been first domesticated 31, 700 years ago, that is a very short time in evolutionary terms. Most of the changes in morphology of the domestic dog to create the very distinct breeds we now see were brought about within the last 150 years. The shape of the animal may have changed drastically in some cases, and certain aspects of the basic behaviour, such as the stalking instinct in collies, have been deliberately enhanced. The dog may have a small brain and little independence compared to its wild ancestors, but the underlying behavioural structure remains that of the wolf. Successful co-existence with dogs depends, to some extent, on the willingness of the human half of the partnership to properly understand the implications and limitations of this on our relationship with them, as is discussed elsewhere in this book.

2
Foundations of Canine Behaviour
David Appleby
and Jolande Pluijmakers

Although the modern dog successfully fulfils the role of companion society places pressure upon it to meet its expectations about its behaviour and yet that same society frequently leaves them ill equipped to achieve these ideals. As a consequence of this failure many dogs are unable to cope with everyday situations without resorting to avoidance or defensive behaviour. The welfare of these dogs is at risk. They cannot relax and enjoy life, they feel threatened by everyday events and are susceptible to stress and disease. As a consequence they are less likely to make rewarding pets and are at risk of being abandoned, re-homed or euthanased.

A review of cases by the Association of Pet Behaviour Counsellors showed that aggression motivated by fear towards people (25%) and dogs (15%) were the most common reasons why dogs were referred for treatment by their vets. Fears and phobias, such as in response to sounds, was the reason why a further 8% of the cases analysed were referred. These figures probably represent the tip of the iceberg because many dogs display problem behaviour related to fear whose owners don't seek help from a behaviour counsellor.

A dog's temperament and character is dependent upon genetic and environmental factors. The extent of the influence of each is not clear, the old 'nature-nurture' argument. This is not the impasse that is often implied because it is safe to say that genetics govern the potentials the dog can achieve and experience and resultant learning shape behaviour and determines whether those potentials are realised. This shaping occurs throughout the dog's lifetime but there is a biologically pre-programmed period in early life during which simple exposure to things or experience of interacting with them has greater influence on the development of emotional states and behaviour than in later life. During this time, called the sensitive period, the puppy learns what to expect and how to react, resulting in a sense of control and the capacity to remain in a positive emotional state in diverse situations. This organisation of a puppy's behaviour forms the foundation on which the development of its subsequent behaviour is based.

Current understanding of the sensitive period has its origin in research conducted in the 1950s, 60s and 70s that suggested that fear of or reducing willingness to approach unfamiliar things develops in identifiable phases between 3 to 12 – 14 weeks of age. However, the interpretation of the results may have been misleading because the researchers depended upon observing changes in behaviour in response to the presence of unfamiliar stimuli. These changes follow the development and integration of the relevant brain structures and a period of learning that results in the organisation of the puppy's behaviour. Therefore the sensitive period may be much shorter than the researchers assumed and looking at the sensitive period from the perspective of emotional development the most important phase is between three to five weeks of age.

Emotional States

It is generally accepted that animals have emotional states such as fear, euphoria, anxiety etc. Whether they experience them consciously remains open to question, although the balance of evidence suggests that they do. Emotions enhance an individual's chances of survival by improving its ability to recognize events that are useful or unsafe and to make appropriate responses. Situations that create a mismatch or conflict with an individual's interests, give rise to negative emotions such as fear that require them to act in order to cope.

Whether an emotion is experienced as negative or positive is dependent on the activity of the dog's autonomic nervous system, which is controlled by the brain. This system is functionally divided into two parts: arousal (Sympathetic) and relaxation (Parasympathetic). These have mutually exclusive actions. Conditions of stress trigger activity in the Sympathetic Autonomic Nervous System that has an emergency function, preparing the body for action. The effect includes increased heart rate, increasing blood supply to the muscles, heart and lungs, increased respiration and bowel movement, sometimes resulting in loss of toilet control, and chemical reactions such as the release of hormones. This activity co-occurs with negative emotional states described in humans as fear, anxiety etc. A positive emotional state (parasympathetic) has a recuperative function due to its opposite effect e.g. by restoring the blood supply and heart rate to normal.

Emotional Homeostasis

Every individual strives to achieve and maintain an emotional homeostasis, i.e. a positive emotional state. The function of emotional homeostasis is to allow an individual to deal with and adjust to the many changes that are part of everyday life. If every change in the environment caused a loss of emotional homeostasis the individual would never be able to relax and its welfare would be seriously impaired. To prevent this, a 'picture' of the world is formed early in life which sets the standard for situations during which they can afford to remain relaxed. Changes in the environment that deviate from this standard may cause a negative emotional state.

Maintenance Set

A dog's capability to remain in emotional homeostasis develops throughout the sensitive period for behavioural organisation; during part of this process mental representations of stimuli are formed and associated with responses. Those associated with a positive emotional state become part of a set of animate and inanimate objects that enable it to maintain emotional homeostasis and organise its behaviour. The effect on subsequent behaviour was shown in three week old Chihuahua puppies fostered individually in litters of four week old kittens (Fox 1971). By twelve weeks of age, the cat-reared puppies preferred contact with cats over contact with other puppies that had not been fostered. Another experiment involved litters of puppies split into three groups. One group was hand-reared from three days old and received no canine contact. The second group was given equal canine and human contact. The third group only experienced other puppies and their dam. When these three groups of puppies were brought together at 12 weeks they showed a preference for puppies that had received similar rearing experience and the puppies raised in isolation from other dogs showed an overall deficit in their relations with other puppies (Fox & Stelzner 1967).

Some maintenance stimuli grow to be so important that the dog becomes dependent on their presence to maintain emotional homeostasis. This dependency can be formed to any social or non-social stimulus, as is illustrated by these experiments. Elliot and Scott (1961) found that puppies confined in a strange pen were three times more distressed than puppies confined in a familiar pen. Cairns and Werboff (1972) housed four week old puppies with adult rabbits and found that within 24 hours of cohabitation, the puppies lost emotional homeostasis when separated from their rabbit, which was expressed by intense distress vocalisations and attempts to escape and regain contact with the removed rabbit.

Dependency on social stimuli is more likely to develop than other stimuli because they provide more stimulation and easily become associated with important factors such as contact comfort, suckling or play. The duration of exposure to a stimulus and its availability are important factors in determining dependency. If a maintenance stimulus, e.g. a female owner, is present in a variety of contexts there is an increased likelihood that the dog will become dependent on her compared to other family members who are only present in some situations. A maintenance set that provides a dog with the best chance of maintaining an emotional homeostasis consists of a large number of stimuli some of which are of comparable worth, which is necessary because the composition of the set alters with changes in the environment and situation the dog is in.

The broader the maintenance set the more able the dog will be to cope with disruption of emotional homeostasis.

Disruption of Homeostasis

Disruption of emotional homeostasis can be caused by the loss of stimuli from the maintenance set, which leads to a feeling of reduced control. The extent to which disruption occurs increases with the importance of the lost stimulus or stimuli. For example, a puppy may whine all night after it is homed because the maintenance set it had built up at the breeders is totally disrupted when it is removed from its dam, littermates and the breeder's house or kennel. Another example is provided by dogs that become so dependent on their owner's presence that they loose behavioural organisation, manifested by an inability to stay in a positive emotional state, when left alone.

Things the dog finds frightening can also cause disruption of emotional homeostasis. These can be learnt if a stimulus becomes associated with a threatening event, such as darkened skies becoming a predictor of fireworks. Dogs may also react fearfully towards a stimulus because of its physical characteristics without having to learn that it is threatening. Examples include intense and/or sudden stimuli as gunshots, fireworks or thunderstorms. As behavioural organization increases, the presence of novel stimuli and stimuli that do not perform to expectation may lead to a negative emotional state and loss of emotional homeostasis.

The Sensitive Period of Behavioural Organisation

The sensitive period of behavioural organisation can be divided into three stages.

INITIAL STAGE

A puppy's initial maintenance set normally consists of its mother, littermates and nest-site. This is unavoidable because of the availability and salience of the stimuli and absence of opportunity to associate other stimuli with a relaxed emotional state because puppies are born deaf and blind and move in a crawl-like manner. Most of their time is spent sleeping or feeding. Although several primitive sensory and behavioural systems are present at birth, the puppy's behaviour is largely reflexive and organized to guarantee what it needs for survival at this stage of development, such as food and adequate warmth. Neonatal puppies are incapable of controlling their body temperature internally but keep it constant by staying in close contact with their mother and littermates and they will display intense distress vocalization when separated from them, but will be content and go to sleep without the comfort of contact with them if they are warm.

Although complex learning cannot occur during the initial stage simple conditioning of touch, taste and scent stimulus-response associations are possible from the first day of life. For example, Fox (1971) placed aniseed oil on the mammary glands of a nursing bitch. Puppies exposed to the smell whilst suckling were more attracted to it away from the suckling context than those that had not had the opportunity to develop the association.

Neonatal experiences seem to have considerable effects on the development of behaviour. This appears to have been illustrated in unpublished work by De Witt, which involved half litters of newborn puppies having no direct exposure to humans whilst the remainder received a high level of human scent for half a minute immediately after birth. Subsequently both groups were isolated from human contact until tested, some at five weeks and some at seven weeks. Puppies that had been exposed to human scent had a preference for investigating people over other environmental stimuli. Unfortunately puppies that had no direct human contact showed the same preference, attributable to the fact that they were mixed with littermates from the exposed group that would have been carrying human scent on their coats. However these combined groups showed significantly greater interest in investigating people when compared to a new control group that was kept in total isolation from human scent until three weeks of age.

A synthetic version of a maintenance stimulus called 'Dog Appeasing Pheromone' (DAP) is available as a plug-in diffuser. The appeasing pheromone is produced by the sebaceous glands in an area between mammary glands of the bitch three to five days after puppies are born and helps to create relaxation and dependency upon her. It is suggested that the resultant positive emotional state enables the puppy to remain relaxed as it explores its environment and encounters new stimuli in the next stage and that this reassuring property persists into adulthood. The use of a diffuser may provide a puppy with a salient maintenance stimulus when it is moved to its new home.

SECOND STAGE (3 TO 5 WEEKS OF AGE)
After the opening of the eyes and around 13 (+3) days and the opening of the ear canal at approximately 18–20 days, a puppy is aware of visual and auditory stimuli. The puppy's movements become increasingly co-ordinated and it develops the ability to run but without the grace of an adult. The ongoing development of cognition, perception and motor abilities means that puppies experience a rapid increase in stimulation and because complex learning is

possible most reflexive behaviour that was displayed automatically without conscious thought is replaced by learnt responses related to environmental stimuli.

During the three to five week period puppies are at an optimum stage of development for establishing a broad maintenance set. This occurs for several reasons; firstly they are less inclined to stay close to the maintenance stimuli they became dependent on in the initial stage (the dam, littermates etc.). In addition to their new mobility this is because what has been described as the seeking system, within the brain, stimulates them to actively search and explore their environment. The second factor is that the brain over produces neurons during early development that will go to waste if they don't get used. Although new ones are formed during learning in later life the capacity to store knowledge of the world and how to cope with it is more efficient during these important weeks. Through exposure to and interaction with their environment puppies store representations of new stimuli and variations in familiar stimuli. The third factor is that during these weeks puppies are in a biologically pre-programmed period of parasympathetic dominance, characterised by factors associated with relaxation, such as low heart rate predisposing them to have a positive emotional state While they are in this relaxed frame of mind, puppies easily associate new stimuli with that emotional state of relaxation and they become part of a broader maintenance set.

At the start of the three and five week period, approach and investigative behaviour is directed equally to novel and familiar objects but most attention is paid to stimuli that change rapidly, such as movement and sounds. As recognition and recall memory develops and maintenance sets are refined attention to familiar things declines. Greater attention is paid to those things that are less familiar. These trigger investigative behaviour, whereas things that are extremely unfamiliar begin to cause some minor hesitation to approach. This can be explained by the fact that when something unfamiliar occurs the puppy's brain compares it with the representations it has stored. A successful match will result in a positive association and the further development of behavioural organisation. Failure to match during this phase of development will normally result in efforts to investigate the stimulus or event.

Preparedness to respond positively to novel stimuli during this stage was shown in a response test conducted as part of an experiment by Freedman et al (1961). In this experiment litters of Cocker Spaniel and Beagle puppies were kept with their littermates and dam but isolated from human company except for specific periods. Pups from each litter were taken from the field for a week of socialization at two, three, seven and nine weeks of age and then returned to the field. During this week indoors pups were played with, tested, and cared for daily in three half-hour periods. How avoidance behaviour develops was evident in daily 10-minute tests of the amount of time a puppy spent in physical contact with a passive, reclining experimenter. The two-week-olds were too immature to do much but sleep, eat or crawl about randomly. However, the three-week-olds were immediately attracted and spent most of the 10-minute period pawing, mouthing, and biting him and his garments; five-week-olds exhibited wariness at first, but they became comparable to three-week-olds before the end of the first play period. In contrast seven-week-olds were frightened and wary of contacting the experimenter over the first two days, while nine-week-old pups exhibited these reactions over the first three days. Puppies tested at 14-weeks that did not receive any exposure to people were fearful and remained so even after several weeks of petting.

Experiments have also been designed to reveal a puppy's response to inanimate objects, for example, puppies housed in conditions devoid of objects were placed in a test area with various articles for just half an hour at five, eight, twelve and sixteen weeks. These puppies were found to increase their exploration as they grew older and developed a preference for more complex things. However, puppies first tested when they were over eight weeks tended to withdraw from the items and those not tested until twelve or sixteen weeks old frequently became catatonic with fear (Fox 1978).

During the three to five week period dependency on maintenance stimuli for emotional homeostasis develops very rapidly and becomes progressively important for behavioural organisation. Search behaviour for missing social or non-social maintenance stimuli starts to appear. Little by little evaluation of contexts also starts to occur. In practice this means that to increase the chance of a puppy accepting unfamiliar stimuli without causing loss of emotional homeostasis novel stimuli should be introduced in a familiar environment so the puppy can rely on its existing maintenance set. It also means that a puppy has to be exposed to a great amount and variety of animate and inanimate stimuli and it will benefit from being regularly moved to locations where unfamiliar types of stimulation and variations of familiar stimuli exist. This is at a time when the puppy is in the care of the breeder and therefore this responsibility falls to them.

THIRD STAGE

After the rapid developmental changes in the brain, nervous system and behaviour in the previous stage, from five to six weeks development continues at a more gradual rate. The 'natural relaxed frame of mind' is lost and replaced by a period of high heart rate and sympathetic dominance from week five to a peak at seven/eight weeks followed by gradual decline.

The increasing maturation and integration of various brain structures during the five to seven week period increases the potential for a reaction to novelty. Things the puppy has not had a chance to develop a positive emotional response to now have an increased likelihood of triggering fear, which may be displayed as either avoidance behaviour, fear or fear aggression.

Although the puppy has established a broader maintenance set and behavioural organization to the known environment, the composition of stimuli in it can change through the availability and importance of other things it encounters. In general things that are moderately different or unpredictable will still cause exploration and a preference for investigating novel stimuli and locations over those that are familiar continues, so long as contextual factors are reassuring. However, the introduction of novelty becomes more likely to upset than to benefit the system. The amount and nature of previous experience increasingly determines whether the puppy shows fearful responses and the capability to control its environment and thus maintain behavioural organization becomes very important. If a dog does not have a sense of control it may easily learn fears or develop what, from our perspective, may seem to be inappropriate and unpredictable behaviour in response to unfamiliar or aversive stimuli. Once developed these responses may generalise to other things and situations.

During this stage puppies should be given plenty opportunity and time to become familiar with new things and build up their confidence to approach and investigate them. Giving them the chance to do this at their own pace allows them to control their emotional state, which they can exercise by governing their approach, withdrawal and wait and see responses. If they stay relaxed the stimulus will eventually be associated with a positive experience. Conversely, forcing a puppy to approach will have the opposite effect because a sense of loss of control will cause anxiety or fear, resulting in a negative association.

To maximise the chance of the puppy forming a suitable maintenance set and coping with the world beyond the breeder's environment puppies should be homed as soon as it is allowed, which in the UK means from six weeks. The younger the puppy the easier it will be for it to add new stimuli it encounters to its maintenance set. After the age of 7 to 8 weeks, when the puppy's representation of the world is, for the most part, formed this becomes increasingly difficult, but definitely not impossible. Obviously, the opposite is true and the later broadening of the maintenance set starts the less likelihood of developing sound temperament there is and starting from scratch at 12 to 14 weeks is likely to present new owners with extreme difficulties. Of course, change is possible but prevention is better than cure. After they are obtained and have received their initial vaccinations the process of exposing puppies to as many different stimuli and environments as possible to build up their confidence should continue. This process has to be repeated until at least six months and probably a year to prevent behavioural organisation regressing.

From 1959 practical application of the research into the effect of early experience was conducted with over 24,000 puppies by Guide Dogs for the Blind Association in the UK. This work is ongoing but was reported on in 1991. The qualities required of a potential guide dog, the training of which starts at around 12 months, include confidence and low reactivity to normal stimuli. Until 1956 the GDBA relied on the donation of adult dogs to maintain their stock and the success rate of those accepted into training and becoming guide dogs fluctuated between nine and eleven percent. In an attempt to improve the success rate puppies were purchased and placed in private homes between ten and twelve weeks of age or even later but it did not significantly improve.

In 1959 the GDBA started managing the gene pool with its own breeding scheme. The puppies were raised in kennels and placed in the urban homes of puppy walkers after first vaccination at six weeks. Systematic exposure to the environment in the area around the puppy walker's home commenced immediately and continued until the puppies were returned for training. Annual success rates between 75 and 80 percent became common with some variation between breeds.

Dogs that exhibit avoidance, or aggressive behaviour related to fear should be more likely to have a history of limited early experience than dogs displaying other types of behaviour problem. This theory was tested by Appleby, Bradshaw & Casey in 2002 when the early experience of 820 dogs referred by veterinary surgeons to one pet behaviour counsellor (D. Appleby) were compared. The results suggest that there is a relationship between a dog's early environment and the development of fear-related behaviour problems. Living in a domestic environment at the breeders and experience of urban life after vaccination appears to reduce the probability that the dog will develop avoidance behaviour or aggression in response to unfamiliar people. Interestingly, there was a significant relationship between the dogs that had been reared in a non-domestic maternal environment, such as a barn, shed or kennel and aggressive behaviour when examined by the vets in later life. A possible explanation is that for puppies that have not had the diversity of experience available in a domestic environment the first experience of visiting the vet, which normally occurs within two days of being obtained, is likely to be traumatic resulting in an association between the vet and fear and defensive behaviour during later consultations.

HOME ALONE

The importance of exposing puppies to unfamiliar stimuli gradually also applies to teaching them to adapt to changes in their maintenance set when left alone. Elliot and Scott (1961) found that puppies that are not exposed to separation experiences early in life have an extreme reaction to them in later life. They divided some puppies into four groups and exposed them on a weekly basis to separation in a strange pen. They repeated this until the age of 12 weeks. Puppies of group four were not exposed to separation experiences until they were tested for the first time at 12 weeks. These puppies panicked and were unable to cope whereas the other groups, especially the group that started having this experience at three weeks, had learned to adjust more effectively. Care should be taken to develop a puppy's capacity to cope with segregation gradually.

Young puppies have a strong need for close social contact. This need is first satisfied by its mother and littermates and later by its owners. Locking a puppy in a room on its first night in its new home, after its previously established maintenance set has been removed, and letting it cry until it is worn out is not good for its welfare. It could sensitise it so that it reacts negatively when left alone on future occasions, increasing the potential for separation problems such as destructiveness, vocalisations and elimination. New owners can gradually accustom their puppy to being on its own by getting it used to an indoor kennel and placing this next to their bed and gradually moving the kennel away from the bedroom once the puppy has learnt to settle at night or they can sleep near the puppy downstairs for the first few nights, gradually withdrawing their presence once the puppy has adapted to its new environment.

Summary

Maturation and integration of structures in the brain makes more detailed information processing possible resulting in changes in behaviour and an increase in emotionality. Once the naturally predisposed period of parasympathetic period of dominance and positive emotional state has declined and the maintenance set has been formed, unfamiliar stimuli encountered may cause a negative emotional state. The characteristics of these stimuli and the characteristics of the maintenance set will determine the extent to which arousal and disruption of behaviour occurs. The more effective the maintenance set the better able to cope with disruption of emotional homeostasis the dog will be. The presence of an effective maintenance set also increases the individual's confidence to explore and broaden it. It follows that a failure to develop an adequate maintenance set during the period of parasympathetic dominance between three to five weeks and beyond should have a detrimental effect on the development of subsequent behaviour the dog's welfare and there is an increase in the potential for behavioural disorders to develop.

3
Puppy Classes –
A League of Their Own!
Erica Peachey

Case Studies

1. Jingo, a medium sized crossbreed was referred to me for aggression towards other dogs. If on his lead, the sight of any dog, big or small, male or female, would cause him to lunge forwards, barking and snarling. Although initially better off his lead, Jingo had now reached the stage where he would race up to another dog if he had the opportunity and immediately grab it around the neck and pin it down. If the other dog fought back, a fight would ensue.

   'I don't understand it' Mrs Smith told me, 'he used to love other dogs. Everyone always laughed at the puppy classes because he never listened to me, he just wanted to play with the other dogs.'

   Jingo had been acquired as a young puppy from a local rescue kennels. Keen to do the right thing, Mrs Smith and her family had enrolled him for puppy classes as soon as he was clear from his vaccinations. Jingo only had eyes for the other dogs, the highlight of his lesson was when all the puppies were let off the lead to 'socialise' with each other. He became the joker of the class, leaping over other puppies, bouncing on them, and generally encouraging them to play. How could a puppy who loved other dogs develop such an aggressive reaction, or was it this same rough play that had caused the aggression?

2. Milly loves people. She has a lovely nature – when she sees a person, all she wants to do is be close to them. Unfortunately, this means that her owner can no longer let her off the lead on a walk because Milly will run over to anyone and jump up repeatedly. A powerful, young Golden Retriever, she has knocked young children over and has even knocked dog-loving adults off balance.

   In the home, Milly's passion for people means that visitors are rare and the children are not able to have any friends to play unless she is shut away.

   'The instructor at the puppy classes said that we should socialise her, so we took her everywhere and she met hundreds of people. She was so cute and small then that no-one minded when she jumped up, some people even encouraged it'

   Was Milly's owner given wrong, or incomplete advice? Would Milly have been a better behaved dog if she had not attended classes?

3. Miss Brown told me that Sheeba had always been the perfect dog. Sheeba had always been shy of other people and dogs and on walks she did not cause any problems, keeping her distance from everyone else and never moving far away from her owner. Then one day, when Sheeba was five years old, she was on her lead on the way to the park, a small boy jumped out from his drive and threw his arms around her. Sheeba struggled to get away, with a yelp or a snap, but unfortunately a tooth

grazed the boy's cheek. It was not a serious injury, and fortunately the boy's parents were understanding, but Sheeba's owner was devastated. Sheeba had not attended any puppy or dog classes. 'I never saw the need' Miss Brown told me. 'She always did everything I asked of her'. Would Sheeba's attendance at puppy classes have meant that her negative response to this incident would have been avoided?

There is no easy answer to these questions. We cannot know because we are dealing with pet dogs in families and therefore cannot run controlled studies to find out what would have happened if things had been different. However, certain patterns are identifiable and this allows us to understand the causes and what programmes of behaviour modification are likely to be able to help.

First Experiences

We all remember our first experiences of something important. Music lovers will know the first record they ever bought, but probably do not know the second. I remember how much a bag of chips used to cost when I first bought one, but have no idea when they went up. How many people remember their first kiss? But what about the second? First experiences are very powerful in our memory and there is no reason to expect this to be any different in the dog. So it is essential to ensure that the first time creates the right habit and the correct learning. In times of stress many of us revert to the first thing we were taught often without realising it. Therefore, if a puppy's first experiences were unpleasant resolving what it learnt can be a long and difficult process. However, good initial experiences will greatly assist a dog through-out its life.

Socialisation and Socialising?

There are many and varied definitions of socialisation and these change according to time and context. We talk about adults 'socialising' during a night out, children 'socialise' when they play together. According to different dictionaries, 'socialise' means:

either:
1. To behave or interact in a sociable manner.
2. To make somebody fit or prepared for life in society
or
1. To make social
2. To act in a sociable manner.

Therefore, in general terms, socialising refers to behaving in a sociable way, or the process of learning to do this.

However, 'socialisation' has a stricter ethological meaning. It can be defined as 'the process during which animals are particularly receptive to stimuli and when various social preferences are established.'

It is commonly thought of as the process of learning how to recognise and correctly react with the species with which it is familiar. For wild animals, such as a wild dog or wolf, this is relatively easy as this involves one species. However, for our pet dogs, at least two species and possibly more are familiar, humans and dogs, and possibly cats or others. Some regard this as socialisation, others do not. We can immediately see that we are expecting more of our pets and that selective breeding has allowed them to cope with this 'extended family' as their wild counterparts never could.

Another definition of socialisation is that 'an animal develops appropriate social behaviour'. Therefore, a dog is said to be socialised to something, when it displays a recognition and acceptance of it and behaves accordingly. It therefore follows that two puppies may experience the same circumstances but because of differences within them, one may be said to be 'well socialised' and one may be 'poorly socialised'.

Because of the popularity of pet dogs, the definition of socialisation has broadened in meaning. Rather than simply referring to the development of an understanding of one or two species, it is commonly used to mean the entire process of reacting to and fitting in with life as a pet dog. In the past, this has been termed 'social referencing' but as discussed in the second chapter – the development of maintenance stimuli to enable behavioural organisation and emotional homeostasis is a more accurate concept. This model is able to cover social and non-social stimuli, i.e. learning to adjust with people, traffic, vacuum cleaners etc. The aim is to equip the puppy or dog with the necessary knowledge to allow it to adjust and behave in an acceptable way in our society. It results in a dog that is better able to cope with novel experiences, unpredictable situations and fearful circumstances. Breed differences play a part in this. Some breeds and types have been selectively bred to react in a certain way. For example, gundogs are normally required to work closely with people and react well to other people and dogs. Livestock guarding dogs are bred to bond closely with their flock and keep intruders away. Terriers have been required to work independently of their owners, pursuing prey and not back down from threats.

So What Does This Mean in Practice?

We want to raise puppies that can cope well in our society. What we are aiming to achieve can be described as: '*an active process of teaching a dog how to behave in all situations, through pleasant experiences.'* This definition encompasses several important points.

- *'Active process'*. This means actively searching for different situations and controlling the circumstances to ensure the right outcome is achieved rather than waiting for things to happen.
- *'Teaching'*. We expect a great deal from our dogs and want them to behave in ways they would not normally do. Therefore, it is unrealistic to expect them to learn it all by themselves. As the humans in the relationship, we must do the teaching.
- *'A dog'*. This applies to a dog throughout his life. Although puppies learn more quickly and have a greater need to learn, dogs are extremely adaptable and it is the fact that dogs are good at learning which enables them to be kept as pets. Dogs of any age can learn but prior learning must be taken into account.
- *'All situations'*. All the circumstances in which a pet dog may find itself. Thought and consideration must be given by owners and those involved in puppy education about what situations can and should be included.
- *'Pleasant experiences'*. This term reflects the fact that dog training has moved on considerably in the last 20 years. The old style of force and punishment is no longer acceptable and has proven to be unnecessary. Deeper understanding of dogs, relationships and learning have meant that reward-based methods that are kind, fair and effective are more commonly used.

There is no set time scale for this process because learning is ongoing. Things that can be learnt can usually be unlearnt and a dog's behaviour will usually change over time. Consequently, active social experiences must form a regular part of life, especially if social behaviour was not well developed in early life.

A litter of wolf puppies which were hand raised, but still had access to other wolves, gradually reverted to their natural suspicious behaviour and were wary of humans even though exposure to humans was constant. As they matured, they chose to keep away from the humans, even those who had hand fed them. A litter that was hand raised away from the company of other wolves, learnt to enjoy people being around and were able to be reintroduced to the pack and still retain their enjoyment of being with people.

Where Do Puppy Classes Fit Into This?

Puppy classes have developed from a tried and tested idea brought to the UK by Ian Dunbar, in the 1980's. Although puppy classes had existed previously, the idea of a class full of young puppies, just after their vaccinations, was a new one. However, like all great ideas, the general theory was sometimes misunderstood or misapplied. Puppy classes – i.e. a class for young puppies to mix and learn through reward oriented methods – gained much popularity and interest and others tried to emulate it. A puppy class can simply mean a beginners class, where young puppies are alongside bouncy adolescents and maybe older dogs, perhaps showing anti-social behaviour. Exposing a puppy to these conditions can cause problems.

The term 'puppy class' does not specify an approach to training. 'Old style dog training', of yank and pull, was originally only for dogs of six months or older. This was because it was thought that puppies under six months were not able to cope with the stresses of this type of training. However, using motivation, lures and fun meant that puppies could begin training from the moment they came into their new homes, thus removing the need for this restriction. Unfortunately, when the 'old style training clubs' sensed competition from forward-thinking trainers, they too reduced their age limit, thus exposing young puppies to choke chain and force.

Everybody has a slightly different concept of a puppy class. It is a good idea for an owner to consider what they want using this chapter as a guide and invest time in finding the right class for them. Each puppy will also have different needs. A good instructor should recognise what they are and cater for them, by providing controlled situations, advice and support for the owner. The nearest puppy class is not necessarily the best, neither is the cheapest. Find out what is available, speak to your veterinary surgeon and other dog owners, talk to the instructor on the telephone and perhaps visit the class to watch, without taking your puppy, preferably before you obtain it.

A puppy class should differ from many conventional training classes in several ways, including:
- Being less structured or regimented. Puppy groups should be small enough to allow the instructor to be familiar with each puppy and family and adapt the course according to their needs.
- Giving more freedom and play which can then be developed into owner control. Teaching about games and the importance of playing the right ones should form a part of classes for any age, but practical exploration of games is possible with puppies.
- Including some off-lead sessions (see later). This does not mean off lead 'free-for-alls', rather structured sessions

which involve off-lead work, such as coming when called, calling away from other puppies, play etc.

- Providing exercises that are varied and imaginative, teaching what is relevant to a pet dog rather than obedience style routines. For example, teaching a puppy learning to come when his owner calls, leaving another puppy it was sniffing, rather than simply returning to his owner after being positioned sitting facing the owner, anticipating the next part of the exercise.

There is a difference in the attitude of owners attending puppy classes. Rather than attending class out of desperation with an adolescent that is causing increasing chaos, owners bring their puppy, who is special to them. Although he may not be perfect, owners are normally far more positive than they may be if they attend classes for the first time when the pet is older. Owners are also at their most motivated and receptive and because results tend to come quickly, they usually try harder.

Puppy classes can only be as good as the instructor who takes them. The term itself does not guarantee a class which will be an ideal learning environment for any puppy. When I use the term for the rest of this chapter, I will be talking about properly run classes where numbers and ages are limited and the instructor has a good working knowledge of owners and puppies learning and motivation.

Some classes are termed 'Puppy Socialisation Classes'. Whatever definition is used, this is somewhat restricting the scope of a class. I feel that puppy classes should actively teach:

- good manners,
- response to owners wishes,
- responsible dog ownership,
- avoiding problems,

- overcoming typical puppy difficulties such as mouthing and any house training problems,
- increasing owner's understanding of their puppy,
- educating the owner about how to shape their puppy's confidence and behaviour in all situations.

The title 'Puppy Classes' more aptly covers this.

Whilst on the subject of terminology, many owners tell me that they and their dog or puppy have taken part in 'The Socialisation Exercise' at local dog clubs. For those not familiar with it, the exercise usually consists of young puppies or beginner dogs or more worryingly, a mixture of both positioned in a circle or line. One at a time, one owner and puppy weaves in and out of the other dogs, usually accompanied by much shouting of 'Leave it!!' and pulling from all the owners. How can this be teaching dogs to be sociable towards each other? It seems far more of an 'anti-socialisation' exercise. From the dog's point of view, stress is inevitable, either from the other dogs being close, possible aggression from other dogs, from not being able to behave in a natural way or from the noise and action of the owners. If we believe that the aim of puppy classes is to increase a puppy's ability to cope, and to teach it how to behave in normal settings, it is hard to see how this exercise is anything other than destructive.

What Should be Included in a Puppy Class?

A good puppy class should be flexible and therefore the content of the course should vary to some extent. Some areas should always be covered, such as puppies learning to respond to their owners, owner understanding, social manners and avoiding problems. Within this, there should be room to cater for the individual needs of the puppies and owners.

Some owners imagine that all the social experiences their puppy needs should be provided by the puppy class. Although a puppy should gain valuable experience, it is unrealistic to think that this is sufficient. Dogs learn according to the situation, so even if a puppy does experience a wide range of people, men with beards, women with hats, children with sunglasses, disabled adults, people from different ethnic groups, etc, there is evidence to suggest that this may only enable the puppy to accept all of them within the confines of the classes. In order for the growing puppy to behave appropriately in all circumstances their education must be continued at home and on walks. Therefore, the role of the classes is to:

- provide basic social experience, i.e. good experiences with different men, women and hopefully children of all ages as well as other puppies.
- to explain to the owners the importance of teaching acceptable behaviour and encourage them to continue and broaden the education between classes.
- to note the reactions of the individual puppies and give each owner specific advice, including areas to concentrate on, times to take extra care and how to avoid problems.

What Else Should Be Included?
FREE PLAY
This is an area of some controversy. The original principles adopted from America involved large amounts of free play for the puppies. However, in practice, in this country, it was found that many puppies were actually learning to fight during these 'play' sessions. Some puppies played so roughly that aggression became part of their game. This was especially true of some breeds such as terriers. Others became

frustrated when their expectations of rough play were not met, which, in later life, can result in aggression. Some puppies were overwhelmed by the boisterous puppies and learned that aggression was the best method of defence. Owners likely to attend a puppy class are also the owners who take their puppies out and about on walks, on which most will meet and interact with other dogs. As a consequence, they are already learning fluency in 'dog language' and this, combined with the artificial, relatively sterile environment of a puppy class, i.e. where there are fewer distractions and the owners are unlikely to disappear by walking away, means that games become the only thing to concentrate on and therefore, frequently get out of hand. In a good puppy class, each puppy will be assessed and the course tailored according to their needs. A shy puppy may be off the lead and encouraged to interact and have good experiences with others, whereas the owner of a boisterous puppy may be shown how to be more interesting than everything else, by using rewards and motivation. The puppies are individuals and different exercises are needed to bring out the best in each of them. This approach must be explained to the owners, to develop their understanding and to ensure that no-one feels 'short changed' or that favouritism is leading to different treatment. Free play sessions, although a superficially easy option for the instructor, are not recommended other than in exceptional circumstances. However, off-lead sessions, when one, two or more puppies are off their leads whilst working with their owners, are achievable in most puppy classes and can be extremely beneficial.

RESTRICTION AND RESTRAINT
Learning how to behave when restricted is very important for puppies and something which is often overlooked by

owners and instructors. If puppies are used to freedom and the ability to interact as they choose, being restricted on a lead is contrary to their expectations and they may learn to object. A gradual introduction to being restricted on a lead is an important step.

APPROPRIATE BEHAVIOUR WITH ADULTS

Teaching social behaviour involves teaching appropriate behaviour. Teaching acceptable behaviour varies between puppies and age groups. A puppy that is nervous of people needs to learn that people are linked with good experiences and that being around them is enjoyable. A boisterous, extrovert puppy that adores everyone does not need to learn to enjoy being with people. It already knows this, but it needs to learn to greet them in an acceptable way. It needs to learn that jumping up results in no attention and keeping feet on the floor is the best way of seeking attention, although this does not guarantee that attention will follow.

MANNERS WITH OTHER DOGS

This must not be overlooked. Puppies will develop their own rules for games and interaction with older dogs which is essential for well-balanced behaviour. A puppy should meet playful dogs as well as dogs that are less inclined to play. They should meet dogs that want to be with them and dogs that do not. Obviously, a puppy should not be around a dog that is aggressive towards puppies but understanding that not all dogs want to play is an important skill for a puppy to learn. If a puppy does not learn to read dog language it is likely to encounter problems as it matures. This should not necessarily be provided during the puppy classes, as finding suitable adult dogs and ensuring they are not stressed is very difficult, but the process must be explained to owners and help and encouragement given at each session.

UNDERSTANDING OF CHILDREN
Manners with, and a basic understanding of, children is something that all puppies should learn. Learning may be limited in a class environment, but it is important to make the effort, perhaps by having one week as 'Children Week', where everyone is asked to bring children if possible, and asking children who attend a different training class to come and meet the puppies. It must be explained to owners that this is not the end of the process and that it is one that must be followed during the weeks ahead. If a puppy lives in a household with no children, every effort should be made to find as wide a variety as possible. If there are children in the family the puppy usually understands them, but children of different ages should be introduced.

HANDS ON
In the old days of forceful training, most dogs became used to rough handling. Although the means did not justify the end result, this tolerance was a useful by-product of the harsh methods. Modern methods, although effective, enjoyable and largely stress-free for puppy and owner, do mean that a puppy receives less hands-on contact. This can cause problems with grooming, administering tablets, being handled by a veterinary surgeon and sometimes even being stroked by a stranger. Therefore, as well as the 'hands free' methods of training which involve luring rather than forcing, a good puppy class should teach the importance of lots of handling, throughout its life. During

the class, this can take the form of physical praise, handling and grooming by the owners and others, with many people touching the puppy in a pleasant way. As with all other aspects of the classes suggestions for practising at home and recommendations for each puppy as an individual are what make the difference between an average class and an outstanding one.

What Does the Owner Need?

Having looked at what the puppies need from a class, we must also consider what the owner requires. This can largely be divided into five main areas:

- *Information and Education.* This is obvious, but the class should be small enough to allow for specific as well as general points to be included.
- *Acceptance* is important. No-one likes to be told off and as adults, very few of us are prepared to give up our money and time for this. Many people still believe in the outdated concept of 'No bad dogs' (remember first learning experiences and the impact they have). This means that the class becomes a daunting place if the owners feel they will be blamed for their puppy's less than perfect behaviour.
- *Support* is reassuring and essential. Some of this comes from the instructor, sometimes it can come from other owners. For example, when the instructor says that mouthing is natural for puppies of this age, it may help, but when another owner confides that their puppy does the behaviour it can put any worries into perspective.
- *Motivation* of adults can be difficult. However much they love their puppies and have chosen to come to classes, we all lead busy lives which sometimes verge on hectic. Motivating adults to change their behaviour and

to spend time on walking, socialising, playing and teaching can be difficult. Consequently the instructor needs to understand human, as well as canine, psychology.

- *Enabling* is vital. 'Give a man a fish and you feed him for a day, teach him to fish and you feed him for life'. Enabling an owner to lure his puppy into a 'sit' position is pleasing for the owner, but teach him how to teach his puppy how to respond to him, and he can teach that puppy anything he chooses. A puppy class only runs for a few weeks, but the skills an owner learns should stand him in good stead for the entire life of his dog.

There are many other aspects of puppy ownership which can, and perhaps should, be included, such as responsible dog ownership. This is not a comprehensive list of what a puppy course should include. For many years there has been an emphasis on teaching response to commands, i.e. a puppy must be able to sit, stand, lie down and stay. This is not included in most people's requirements. Practical exercises such as coming when called in a natural setting, walking on a loose lead and greeting people without jumping up are much more likely to be listed.

Puppy Parties

Puppy Parties can greatly benefit a puppy, owner, and their veterinary practice. The definition can vary, but a 'puppy party' should mean a group of very young puppies and owners being invited to a veterinary surgery for a one-off meeting or a short series of meetings. The puppies are normally very young, usually between first and second vaccination. The group should be small, up to about eight puppies, of different breeds and sizes but of similar age. Whole families should attend, providing a variety of learning experiences for the puppies. During a puppy party, the puppies and owners receive an introduction to the veterinary practice and form pleasant associations with it. They are also introduced to the concept of socialising and other behavioural organisation and to training and understanding their dogs. Avoiding problems and resolving common difficulties such as mouthing, lack of house training etc should also be included. As the puppies are so young, more 'off-lead' time is usually appropriate. This is an age when puppies are usually isolated from many of the situations because the vaccination process and experiences which they so desperately need. A puppy party goes some way towards compensating for this confinement. Owners are usually extremely receptive at this stage and therefore can enjoy the puppy party and be motivated to continue to provide social experiences and learning. There are risks to puppies mixing before their vaccination programme is completed, but at puppy parties these are minimised and therefore, it is as safe as possible. Although hard work for the veterinary practice to organise and run, the puppy party yields huge benefits for the practice and its clients.

Conclusion

So where do we go from here? It depends on what we are looking for: for those involved in research there is a wealth of opportunity to study puppies in pet homes and to monitor the depth of quality experience and its effects.

For those involved in puppy education, we constantly strive for improvement. With the advances that have occurred in training methods, instructor courses and support, knowledge is now easier to find but maybe harder to keep up with and to put into perspective as useful tools.

For those with puppies, decide what you want and find

ways to achieve it, by 'wearing your puppy' and taking him out and about with you. Hopefully, a good puppy class will help you achieve your aims, understand more about your puppy and have fun in the process.

Think back to the three dogs we met at the beginning of the chapter. Have your views on them altered in any way?

4
Children and Pets
Erica Peachey

Introduction

Pets and children together, the happy ideal which we have in our minds. Whether it is on old fashioned chocolate boxes, Lassie films or advertisements for toilet rolls, the idea of adorable children and cute animals can be a far cry from reality.

For both pets and children being involved with each other has tremendous advantages. It is undoubtedly hard work for the adults involved, but it is worth the effort. For example, studies have shown that children raised in households with pets tend to relate more closely to others, have greater empathy, increased social skills and may have a better understanding of life's difficulties. For pets, especially dogs and cats, the rewards can include a richer, more varied lifestyle, increased companionship and time spent with them. Sometimes, I wonder why I have found no studies outlining the benefits to parents. On difficult days with my family, I can imagine why this is, but most of the time, the rewards are obvious.

Dogs pose the biggest potential for problems with children but also, perhaps, the greatest opportunity for good times, so the majority of this chapter will focus on this. It is divided into four sections. The first looks at dogs and children in the same household, growing up together with all the difficulties this entails. The second section looks at some of the special considerations which may be needed. Thirdly, we look at advice for dog owners and finally, advice for parents. The information is aimed at the majority of dogs and children, although there are exceptions to every rule.

Know your own dog and/or child and follow, adapt or disregard advice accordingly. If you have any concerns regarding the behaviour of a pet, speak to your veterinary surgeon and ask for a referral to a reputable behaviour counsellor, such as a member of the APBC.

The general theory is as follows:

1. The aim is to provide good experiences between the child and the pet.

   What constitutes a 'good experience' will vary between different children and pets. For dogs, this will tend to include walks, games, food treats and attention. It is preferable that the dog has learnt good manners and the boundaries of acceptable behaviour before being with children. For example, if a dog snatches treats, barks constantly for toys, jumps up when going for a walk etc, pet ownership is unlikely to be a good experience for the child involved.

2. As well as good experiences, it is essential to reduce or stop bad experiences

This may involve:
- Limiting the time the animal is with the child. This is especially important if the child is active or excited.
- Not allowing the child to chase a dog that does not want

to play. What is the child learning? What is the dog learning? How can they learn to like each other's company from that experience?

- Not allowing children to tease/hurt/pester the pet. This may sound obvious, but it is amazing how often it is overlooked.
- Not letting the pet frighten the child. Barking, jumping up and licking may all be perceived as friendly, but can be very intimidating for a young child. Therefore, take special care around greeting and games, and especially with puppies who have not learnt to stop mouthing.

3. Other points

A child's presence should signal good things. Find ways of including the child in the pet's favourite things. For example:

- walks with the baby and Rover,
- games with the child involved or sitting on a knee,
- cuddles with everyone involved etc.

It is tempting to wait until the child has gone to bed and then have quality time with Rover, but what does the pet learn? That life is great as soon as the child is out of the way? Added to which, the child cannot learn about interacting with the pet. Some things are better done when the children are not around but ensure that good things happen when together. When the child goes out or to bed have a period of time of no interaction with the pet, so it does not learn that no child equals fun and attention.

Routine. Routine is often helpful for pet, child and adult but flexibility is also important.

4. The three Ss.

Safety. Safety must always come first. Every parent and every owner must be responsible for their own child/pet. If a situation leads anyone to feel uncomfortable, stop, separate and review it.

Supervise. Supervision varies. For example: the advice is that children should be supervised when in the bath. However, this level of supervision varies, from a new-born which has to be held and the parent cannot take their eyes off them, to being in the same room, to being in vocal contact, until the child is old enough to be left when in the bath. Therefore, merely being in the same room may not be an adequate level of supervision, especially in the early stages of resolving a problem.

Common Sense. It has been said that the problem with common sense is that it is not all that common. Lots of information is available, especially as many people now have access to the internet. However, not all information is correct or applicable in each situation, so use your own best judgement at all times and seek qualified professional help where necessary.

Dogs and Children Living Together

WHEN A DOG IS ALREADY IN THE FAMILY AND A BABY ARRIVES

In some ways, this is one of the easier situations, as there is normally a few months in which to make any changes. Thought and consideration is needed to ensure that all remains harmonious. A dog that walks nicely on the lead, does not jump up or demand attention and that comes when called is greatly appreciated when you have a young baby to consider. Before the arrival of the baby any behaviour problems and training difficulties should be resolved. A behaviour modification course supervised by a behaviour counsellor and/or a dog training course are better undertaken now while there is more time and less complications. A thorough check up and chat with your veterinary surgeon can put your mind at rest concerning any health worries (see below).

BEFORE THE ARRIVAL OF A BABY

Decide what changes will need to be made, make them and stick to them. For example, if Rover will not be allowed on the furniture when the baby is born, make these changes as early as possible so that the pet is used to them before the baby arrives. Not only does this mean that changes are made gradually but the pet is less likely to view the baby as a limiting presence. It is also important to accustom the dog to the sights, smells and routines that will occur when the baby arrives. This can be achieved through making a tape of a baby crying, introducing scent impregnated cloths and practising the routines associated with baby care, perhaps with a large doll. Your midwife may be able to help you obtain tapes and cloths from other parents.

PROOF FOR ROUGHER HANDLING

Children can be heavy handed and have to learn how to avoid hurting others. However gentle the child and however carefully you supervise, the child will behave inconsistently, touch in different ways and generally prove to be something of a challenge for your pet to cope with. Try to anticipate each stage of your child's development, and prepare your dog for it. Grabbing fur, hugging, tripping over, shouting, screaming, banging doors, pointing to ears, eyes, staring, jumping off furniture etc all form part of most children's repertoire at some stage and so help your pet, get used to these. For example we want Rover to feel that it is perfectly normal for people to throw themselves on the floor, so that when the child does it, it is not an automatic signal to play. At a time when your child is not around (so they cannot copy) do a minor version of some of the things your child may do. Reward Rover for no reaction. Then do it slightly more, and reward Rover again and so on.

Children copy adults, so ensure that the way you behave with your pets is something you want your child to emulate. For example, if you frequently shout at Rover and tell him he is a 'bad dog', do not be surprised if your child reacts to him the same way. Even if it is a pleasant action such as patting their version of it may not be enjoyed quite the same way. Try to stay one step ahead by thinking of the things your dog will have to cope with and prepare it for them.

Consider any additional points that your pet may need to learn. Acceptance of being occasionally shut out of a room is essential in my opinion. Teaching a dog not to touch toys left on the floor and to ignore the sound of squeaky and rattly toys by using positive reinforcement is also a good plan.

Develop a routine. It is tempting to spend all your time with the dog, especially during maternity leave, but this may not be the best for the dog because it may develop the wrong expectations that will be at odds with reality after the baby arrives. Exercise and games should continue, with as little disruption as possible. On a practical note it may be beneficial to speak to a professional groomer, even if your dog does not normally visit one. Having a dog bathed and excess hair trimmed can be a bonus. Poop scooping can become more difficult for pregnant ladies, but is still essential part of walking a dog.

Decide on what provision will be made for your dog when you go into hospital. Most new mothers and babies stay in hospital for a few days after the birth and it is a good idea for someone to bring some items of clothing home for the dog to smell before washing. The clothes just have to be left lying around for the scents to become familiar.

On returning home, it is usually a good idea for the mum to leave her baby outside (with someone to super-vise) whilst she enters the home and says 'hello' to the dog. Then the baby can be carried in (car seats are wonderful for this) and the dog can say his 'hello' in whatever way is appropriate. If you are concerned about the reaction of a boisterous dog, attach a trailing lead to the dog's non-slip eg buckled collar and this way you can interrupt any over-friendly behaviour as well as rewarding calmness.

New babies are relatively easy for most dogs to accept, if there are going to be major difficulties they are more likely to occur when the baby becomes mobile. Suddenly, this strange creature is moving and unpredictable. At this stage, protect the dog to whatever extent is necessary and begin to teach the child. Although they are extremely young and cannot retain all information, they are learning huge amounts and need to develop good habits. Good experiences together are essential and this normally involves walks, games, titbits etc and maybe cuddle times for parent, child and dog. Ensure that the dog has time away from the child and can always get to a safe area where it will not be followed by the child; baby gates, play pens and indoor kennels can all be useful here, if used cor-rectly. Meal times can be difficult. I find that a good general rule is that dogs and children are fed separately and do not mix at these times.

WHEN A PUPPY IS ADDED TO A FAMILY
Any difficulties can be greatly reduced by careful thought and consideration regarding the choice of puppy including breed characteristics. Ensure the puppy (or older dog) is

acquired from a reputable breeder or home. I would recommend a puppy with an experience of home life preferably with children in his early weeks as a pet for a family.

If acquiring an older dog, ask why the dog needs a new home. Even if the situation is not due to behaviour problems, for example, a breeder selling on older puppies, I would hesitate to take on an older dog that has not been successfully integrated with children. This does not necessarily mean living with them but the dog should have been with children frequently and enjoyed the experience.

All the points discussed in respect of introducing a baby apply, such as good experiences together, separating at times and teaching all members of the family, pet included, how to behave with each other. Good puppy or dog classes will be happy for the whole family to attend to develop this learning, depending on the ages, number and behaviour of children.

Special Considerations

- Some older dogs cope with the family changes when a baby arrives extremely well, enjoying the extra interest that life with a baby brings. Older dogs are generally well behaved and usually need less exercise, socialising and training than when younger. However, extra care may be needed as the older dog may be fixed in a routine and its expectations may take time to change. Also an older dog may find it difficult to get out of the way of boisterous toddlers and may find rougher handling painful and therefore needs to be more protection from the child's attentions.
- It is likely that a younger dog is more demanding of your time as it needs greater exercise and stimulation.

The advantage is that it can adapt easily but you must still give him the time he needed for stimulation, exercise, training and social learning.

• Children with special needs can benefit from being around pets but obviously especial care must be taken that the pet is suitable. There are no hard and fast rules as every situation varies but always see the situation rationally and from everyone's viewpoint, including the dog's! Take advice where necessary and do not rush into making decisions.

• It is necessary to properly approach a situation where children regularly visit a dog, such as is frequently the case with grandparents or other relatives who own a dog. In some ways, this is easier as the dog doesn't have to live with the child. However, it can be more difficult because the dog doesn't have the same opportunities to learn as a dog a child lives with. The frequency and duration of visits is an important factor. At one extreme is the close family, where the child and adults call in regularly, possibly unannounced, for varying lengths of time. In many ways, this is good as the dog is used to the child at every stage of development and they are familiar with it. However, if the dog is not happy with the situation, it can be problematical although the same considerations apply as those for, i.e. supervision, limiting the time together and ensuring both have good experiences.

At the other extreme is the family who visit and stay infrequently. They typically live further away and therefore, their visit usually involves one or more overnight stays and a disruption of routine. In some ways, this is more difficult, as the dog may only see the child every 6 or 12 months and so cannot be easily accustomed to the changes. The child too has not seen the dog and therefore, has less idea of how to behave with it during these visits. If dog and child really cannot mix, segregation or a good boarding kennels is an option.

• It is essential that the pet should not become a scapegoat for family tensions. Families are rarely harmonious all the time, but if there are unresolved issues, the pet should not be involved.

Meeting Unknown Children – Advice for Dog Owners
In these days of court cases and litigation, children can strike fear into the heart of a dog owner. What can be done to reduce the likelihood of any problems?

Socialising the dog is vital, and this should continue throughout a dog's life. Although patterns of social behaviour become established over the first few months of life, dogs are always learning and like a person driving a car, it is better to practise these skills on a regular basis. The concept of behavioural organisation (as discussed in Chapter 2) in part, involves learning the right behaviour through pleasant experiences. Therefore, socialising to children should not be a matter of leaving a puppy with some children and hoping they all learn to get on or mixing with a couple of children and hoping this is sufficient. It is an active process involving seeking children of different ages and types and ensuring that your dog has a good time with them.

Socialising with children can be difficult, especially if the dog does not live in a household where there are children. Good puppy and dog classes provide the presence of some children but more will be needed. Walks in different places, mixing with friend's families and walking in busy areas in a controlled way can all help to increase the amount of contact with children. Be ready to make the most of any situation. For example if a child or a parent shows any interest in your dog, ask if they would like to make friends.

Anticipate and avoid difficult situations. If you feel unsure of the child's or your dog's behaviour, walk away from situations you are in, removing your dog from potential trouble. This may be because a child is becoming too boisterous, a small number of children has becomes a large group or because you simply notice signals from your dog that it is no longer enjoying the contact with the child or children and is beginning to find it difficult.

Some children may be scared of dogs in general or of your dog in particular. Children who are scared can behave in a way which may dogs find unnerving, e.g. if the child tends to stare, squeal, move quickly, comes close and then moves back. Know your own dog and do not allow a situation to develop it cannot cope with.

Some children are too rough. These tend to be those who are unused to dogs and therefore have no idea of how to behave around them, or children who live with a very docile and tolerant dog and have learnt that the way to interact with dogs is to pull their ears, hang from their necks etc. Again, remain in control of the situation and be ready to interrupt them and give some guidance or move your dog away if needed.

The attitude of parents is also worth mentioning. It came as a great shock to me when I realised how many adults are afraid of dogs to a greater or lesser extent and this can have a negative effect on their children. Make sure that you don't force your dog on anyone and just as you and your dog may need to 'escape' when you choose, the parent and child must have the same opportunity. A common misconception is that only aggressive dogs cause problems. This is not the case – many problems arise with friendly dogs, especially when they are off lead. A dog that loves children can be an extremely frightening creature to a child who is not confident around strange dogs. Being knocked over by a dog is very frightening and many playful dogs become extremely excited if this happens. Take extra care if you feel this may apply to your dog. You may know that your dog is friendly, but behaviours that a dog owner may view as normal can be enough to arouse feelings of fear in parents and/or children. This will include barking, running to greet people, jumping up, being playful, running quickly, licking (especially faces), puppy mouthing etc. If your dog has a tendency towards any of these, you are likely to encounter hostility in public places but appropriate training and control can prevent these problems.

Always remember that your perception of your dog may not be the same as other people. Many children are frightened of large dogs, but assume all small dogs are puppies and therefore want to be cuddled. Certain types arouse feelings in adults. As a generalisation, large dogs, black dogs, German Shepherds, Rottweilers and Staffordshire types appear to be viewed with greater suspicion and less tolerance than many other breeds. For example, owners of black Labradors tell me they meet more anti dog feeling than is reported by owners of yellow Labradors. Due to reduced tolerance the presence of dogs around schools and children's play areas is increasingly frowned upon. This is such as shame because responsible dog owners and well-behaved, friendly dogs can be a source of great interest, education and fun for children, some of whom may not otherwise have much contact with them. When out and about with your dog in areas populated by children always take extra care and ensure that your dog, whilst learning his own social skills, helps others to view dogs in a positive way. One final word on being an ambassador for dogs: never forget your poop scoop. Carry it with you and use it when necessary.

In summary, be a responsible dog owner. Each time you are out you are contributing to people's view of dogs. Make sure that no-one has cause to fear or worry about your dog or its actions and if possible, let children meet him and learn a little more about the enjoyment of dogs. Socialise your dog with children and be one step ahead in each situation, anticipating any possibility of problems and ensuring it does not happen.

Meeting Unknown Pets – Advice for Parents

No-one wants their child to be hurt, and being bitten by a dog is a source of fear for many parents. Usually this fear is based on lack of knowledge, or perhaps some experience in their childhood.

Teaching children to be safe around unfamiliar dogs is as important as teaching them how to behave around cars. What we teach and what we expect depends on the age of the child and therefore learning must be developed as the child grows. Even within the same age group, children need to learn different things. A child who is scared of dogs needs to learn confidence as well as basic manners around dogs. A child who is confident with dogs needs to learn to respect dogs they are not familiar with and to read their body language, as well as ways of mixing with dogs which are safe and enjoyable. Being scared and being over confident are opposite ends of the scale, but children at either extreme are at an increased risk of being bitten because of the way they behave.

The rules I teach include:

- Not all dogs are like yours at home. However they behave with their own dog at home (or a relative's dog etc), children must learn how to relate to unfamiliar dogs.
- Do not stroke a dog unless you know the dog's name. This is a useful guide for young children.

As the children get older, I recommend:

- Firstly ask the owner if you can stroke their dog, or give him a titbit etc.
- If the owner says it is fine, you should then 'ask the dog' by standing a little way from him, speaking to him and holding out your hand.
- If the dog approaches in a friendly way, you are fine to gently say hello, preferably by stroking on the chest or shoulders.
- Never put your face close to strange dogs.
- Do not make excited gestures or noises.
- Never attempt to make friends with a dog that does not have an owner with it. This includes dogs tied up outside shops as well as dogs running loose. A dog may look just like a friend's dog, you may think you know it, but still ignore it until an owner is with him.

If a child is unsure of a dog they should be taught to stand still, not scream or wave arms around, just to 'stand like a soldier'. It is best not to stare at the dog, but this is hard to do, so I often teach children that after standing still it is a good idea to slowly look around to find an adult. They should call to them and when they walk away, they should move slowly. If the child is holding a toy or food, I recommend that they drop it and back slowly away.

A Note about Health Issues

Health issues are always mentioned when discussing children and pets. It is beyond scope of this chapter to cover it. If you have any concerns about sick animals, consult your veterinary surgeon. I would advise a visit to the veterinary surgery as a chat with a veterinary surgeon will put any health worries into perspective. Appropriate treatments, vaccinations, fleas and worming at regular intervals, plus regular check ups and will keep most problems at bay. Teach children basic hygiene rules, such as washing hands after touching pets and before eating any food and not to put faces near to pets. Take care with faeces, poop scoop and dispose of it effectively.

Conclusion

The subject of pets and children is huge and one which could be written about in far greater depth. The aim should be to avoid problems where possible by careful thought, socialising and avoiding difficult situations. Where a problem exists or where there is a potential problem, a veterinary surgeon will be able to help by providing a referral to a recognised behaviour counsellor, such as a member of the APBC. Growing up with a pet for companionship should be a wonderful experience, with many benefits. Through careful thought and consideration, good points will be maximised and any possible negative points minimised.

One final note . . .

Whilst caution and common sense must always be exercised, most problems can be resolved or at least improved. Unfortunately, all too often, the general advice if a dog shows any aggression towards children is to part with the dog. This is usually based on concern for the child and a lack of knowledge from the person giving the advice. Although great care must be taken, many problems involving aggression towards children can be overcome by following the right recommendations, and therefore avoid making hasty decisions which cannot be reversed.

Children and Cats

Donna Brander

The first thing to understand about getting any pet for a child is that all situations involving young children and pets are potentially dangerous. Interactions between pets and young children should always be supervised and every effort should be made to avoid any risk to both parties. Although the incidence of dog bites in the population has been well documented, injuries inflicted on children by the domestic cat are not often reported. Nevertheless, being attacked by a cat can be both physically painful and profoundly disturbing emotionally to most children.

Some children are excellent with cats but kittens are fragile and easily injured. Kittens need to be handled properly in the early weeks of their life or they can develop some difficult behavioural problems. Cats and kittens are widely available and inexpensive and it is very easy to get either on an impulse; they may appear to be the easy choice for the family as they are perceived to be more independent and less of a commitment for the child or the family. This is partially true but can be part of the problem. Cats and children both want life on their terms and this can set up a conflict between them.

It is important to understand that cats can be very

subtle in their signalling. Initial warnings can include mild skin rippling, flattening of the ears and turning away and ignoring overtures from people. Children do not always understand a cat's warning signals and this can lead to defensive aggression. A mildly irritated or anxious cat may lay with its paws on the floor (rather than tucked). It may flick its tail as a sign of irritation. This particular behaviour should not be mistaken as friendly tail wagging! Children should be taught the cat is saying it wishes to be left alone.

The signals can become more overt such as hissing, hair erection, and arching of the back, and if these subtle signals are ignored, the response can be swift. Most cats will attempt to move away at this point but some will resort to defensive aggression, particularly if confined or restrained. If the initial, subtle signalling is repeatedly ignored in interactions, the cat may rapidly resort to the more overt signalling. A child often becomes a victim of this behaviour because of a lack of knowledge of what the cat is communicating and how to respond.

Cats can remain reactive long after the triggering event. If a cat has had an unnerving meeting with another cat, dog or human, or even been frightened by noises, it may still be upset and aggressively responsive for some time. If a child attempts to handle the cat while it is still reactive, the cat may respond with what appears to be unprovoked aggression. It is best for the child to always initiate contact or play using a toy rather than touching it by hand. If the cat is overly reactive, it gives it the chance to attack the toy, or run away, rather than frightening the child.

Some cats may respond to unwanted handling and interaction by running away and hiding. This may solve the problem of aggression, but is unlikely to satisfy the point of getting the animal as a pet in the first place. Animals should be brought into the home to teach children to respect and understand them. Having a cat that runs away and hides, avoiding all interaction, hardly satisfies these criteria.

It is fundamental to consider whether a cat is an appropriate companion for your particular child. Choices are frequently made according to the parent's own desires and far too often the child is then made responsible for the animal. Consideration should be given to the temperament and capabilities of the child and the demands they may make on the animal. If the animal chosen doesn't fit the demands and temperament of the child, the child may lose interest or, even worst, may grow impatient with the cat. The family should not have unrealistic expectations of the tolerance levels of a cat. All cats are individuals and have different tolerance levels and most cats can be defensively aggressive if they feel that no other choice is available.

When choosing a kitten for a child, look for a kitten that is responsive, energetic and curious. Shy kittens usually grow up to be shy cats and are not suitable for children. Older cats may be a better choice for children under seven or eight years of age. Check to see if the older cat has a calm response to noise, activity, and friendly overtures from children. Give the older cat time to settle into the home and allow him to make his own friendly overtures rather than having affection forced upon him. Give the cat control by teaching the children to leave the cat alone unless he comes to them. Point out friendly behaviour from a cat e.g. head rubbing or tail in air, that means it is ready to be petted. Teach by example how cats like to be stroked (gently in the direction the hair lies) and how they like to have their cheeks rubbed.

You should make sure the cat has gentle playtimes and also be aware of inadvertently teaching a kitten inappropriate behaviour. Bring up your kitten as you mean for it to go on. It is very cute for a small kitten to attack your ankles but not so cute in a grown cat. Hands and feet should never be used as playthings and rough play should be discouraged between children and cats. Have a variety of toys available for the child and the kitten or cat to interact with.

Be aware! Cats require a lot of rest, in fact, most domestic cats sleep between fourteen and eighteen hours a day! Kittens have short, intensive play/eating periods with long periods of sleep. Some kittens have suffered malnutrition and even death because they were too tired to eat properly after they were handled and played with for too long. Kittens and cats need frequent rest periods all day long with food and water readily available when they are awake.

In some cases it may help to facilitate your cat's introduction to children by linking highly rewarding food with their presence as well as playing with toys. Dogs are rather omnivorous and soon link children with the bits of food they regularly drop. If you want your cat to link food rewards with children make sure it is food of high value to the cat, such as bits of special chicken-meat the cat only associates being with the children.

Give the cat some control over its environment by providing boltholes that are not accessible to the child. These can be high places with box beds and blankets and food and water out of reach of the children. The addition of hidden places under a bed or in a closet can be beneficial. These resting places should be off limits to the children.

This is to allow the cat the space and privacy it needs and teaches the children to set limits to their behaviour in order to respect the rights of others.

As with dogs introducing a cat to an infant can begin long before the baby arrives. Try to get a tape of an infant baby crying and, perhaps, a dirty nappy! Begin to de-sensitise your cat to these novel items. You can even get a doll that moves and cries and give your cat a lot of positive attention, play and food rewards for being curious but calm around the 'baby'.

Make sure litter trays are private places for the cat. Many cats that ultimately have problems of eliminating out of the litter tray have been disturbed while they were toileting in the tray. One unsupervised child was known to pick up and scold the family kitten when he caught it toileting in the litter tray. Needless to say, the cat began toileting behind the sofa and in the closet – anywhere hidden rather than in the tray!

A cat or kitten added to the family should definitely be seen as a family project rather than another playmate for a child. Cats should never be brought into the home to be living toys for the pleasure of the children and a cat may not be the right choice for children who have difficulty controlling their impulses.

Young children should be taught to think of having a cat as a privilege to look forward to when they are 'old enough', rather than as a right of childhood. Having a cat in the family should be about learning about a cat's needs and behaviour, how it communicates, how the child can fit into the cat's world and how to communicate appropriately with it. It should be about respect for the cat and learning responsibility for the care and welfare of another living animal.

Children and rabbits, guinea pigs and small rodents

Emma Magnus

It is very common for children to be given a pet rabbit, guinea pig, hamster or gerbil in exchange for the dog or cat they really crave. The feeling amongst most parents is that these animals represent an easier alternative, a pet that will involve less care and expense. In addition they are readily available at most pet shops and can therefore be purchased on impulse with very little forward planning.

Unfortunately for many children the relationship falls at the first hurdle with the pet becoming aggressive or simply 'not doing anything'. With a gerbil or hamster living for at least two years and a pet rabbit living for anything from five to ten years the purchase of these animals, especially for a child who may lose interest, should be as a result of an informed decision. Most of all, as already mentoned, it is important that pets are not purchased as a live 'toy' for a child's entertainment. Unless they are committed to providing for its needs and maintaining its welfare, albeit with parental help and guidance, it is better that they do not have a pet at all.

The husbandry requirements for each species of smaller pet are unique and care must be taken to ensure that they are being provided with the correct diet and the appropriate environment. For example, guinea pigs are unable to produce their own Vitamin C and therefore require special diets. A gerbil will become very destructive in a shallow cage with no room to burrow, it will become much more active in an aquarium with thick bedding to burrow in. A rabbit requires lots of daily exercise and social contact and may benefit from living indoors where it will be able to run around.

Rabbits, guinea pigs, rats, mice, gerbils and some species of hamster are very sociable and in the right situations a good relationship can develop between the child and the animal but these are all prey animals and therefore have a very different perspective of the owner-pet relationship when compared with the cat or the dog. The interaction between a human and a prey animal involves an enormous amount of trust that can very easily descend into distrust and fear.

The greatest issue affecting a prey animal's relationship with a child is handling. For each species of rabbit, guinea pig or rodent there are a variety of breeds, differing in size and husbandry needs. Some of the larger rabbits are too heavy for a child to pick up; some of the smaller breeds of hamster are too small for hands that may squeeze too tightly. Whichever breed is chosen, it is important to get the animal when it is young, from a breeder that has ensured it has been handled daily since it was weaned. The breeder, veterinary surgeon or another experienced individual should advise on the correct method of handling for the species. An easy mistake made by many pet owners is to pick a rabbit up with hands either side of the ribcage and the rear end dangling. This method is likely to make the rabbit, a prey species now swooped upon from above, feel unsafe which may lead to some kicking and struggling. This is not only dangerous for an animal with a very fragile spine but can also lead them to avoid handling in the future. Similarly smaller pets have to be handled in a manner that is appropriate for their physical characteristics and their ethology. Once parents have researched what these are for the species concerned, they must ensure that their children fully understand and behave accordingly.

An animal that has not been handled regularly, or has

experienced inept handling, is likely to display fearful behaviour each time the owner tries to pick it up, or even tries to feed it. This fear behaviour may be exhibited as cowering and hiding but it may also become aggressive behaviour. For a child, this is hugely disappointing and frightening and will generally lead them to avoid all contact with the animal in the future. For the rabbit this means life at the end of the garden in a hutch, being ignored.

5

The Importance of Positive Reinforcement

Inga MacKellar

Introduction

We live in a technology driven age where people's expectations are demanding and instant solutions and responses seem mandatory. We expect fast, instant food; get annoyed when we get stuck in a traffic jam; become irritated at the supermarket checkout by the person who takes forever counting out their money; we get cross with the dog when it's 'dawdling along', sniffing at every lamppost on its walk. I also challenge any computer user to deny that at some stage they have not become frustrated with it for being too slow! Countless consumer programmes encourage us to complain about poor service and goods and to generally be less tolerant. This demanding lifestyle will invariably have a toll on our pets' lives – be it that we have less time for them or that we have high expectations and want them to respond to our wishes instantly.

There are many types of pet, but the dog remains one of the most popular and the species most frequently referred for behaviour problems *(APBC)*. If we are to have a mutually enjoyable co-existence with our dogs, it is important that we make time for them, make an effort to understand them and think about how we interact with them.

I have found that many owners often resort to punishing their dogs for actions that they perceived to be inappropriate or unwanted, or perhaps for not obeying a specific command. In many instances, this was a result of a lack of understanding of dog behaviour and frustration. Due to the fact that dogs are frequently treated as humans *(anthropomorphism)*, many owners use 'human' rearing and disciplining methods.

In a study of dog owners in the Netherlands, owners were asked how they would react if their dog displayed an innapropriate behaviour. The problem behaviours cited were categorized into eight situations: fearful, aggressive/dominant, 'mating', emotional, bored/lonely, eating, unruly and disobedient. It was found that in most of these situations owners demanded compliance and reacted by addressing the dog severely. Punishing the dog was specified as the first option to use when the dog misbehaves and many of the situations elicited feelings of irritation and anger in the owners together with disappointment and powerlessness (Ben-Michael). Owners frequently tell me *'He knows that he's done wrong as he looks guilty'* and often feel hurt and let down that their much loved and cosseted dog is 'knowingly' disobedient and defying them. On further investigation these owners have, invariably, punished their dog for what they regard as 'bad' behaviour. They would automatically smack a dog if it had urinated in the house, shout at it to stop jumping up or, thankfully more rarely, give it an electric shock from a special collar to stop it wandering out of the garden. One of the most common punishments is reprimanding the dog on returning to its owner after it has 'run away' on a walk.

Repeated use of punishment has not resulted in any improvement in behaviour and, in some instances, had resulted in problem behaviour getting worse. In most of these cases, considerable improvements were made after the implementation of positive reinforcement coupled with *negative* punishment.

How Dogs Learn

In order to understand how a dog might respond in a given situation, it is necessary to have some knowledge of how it learns. In the 1920's a Russian scientist, Ivan Pavlov, found that he could make dogs salivate even though food was not always present. He achieved this by letting the dogs hear the sound of a metronome just before they were fed. Over time, the dogs would salivate to the sound of the metronome even if the food did not appear. He could keep this up indefinitely as long as he sometimes gave food after the sound (Pavlov). This discovery is the basis of *classical conditioning* i.e. a stimulus *(e.g. the metronome)*, that has no apparent importance in itself, becomes significant because it predicts that something will follow *(e.g. the food)* and triggers an involuntary action *(e.g. salivation)*. Classical conditioning can be used to trigger a response over which an animal has no control, including the process involved in emotional states. For example, if a dog gets excited about going for a walk, but is never taken on a lead, then the lead will have no significance and will not trigger excitement if it is produced. However, once a dog has been taken for a walk on a lead a few times. The sight or sound of it will trigger excitement in the dog, even if it is not taken for a walk.

Dogs also learn through *operant conditioning*. In this instance a dog will learn that if it performs an action, that action will have a particular result. Simply put, the dog learns *'If I do this . . . this happens'*. There are four possible outcomes: positive reinforcement, negative reinforcement, positive punishment and negative punishment. Where the outcome of the action is something pleasant or beneficial, the dog is *more likely* to repeat that behaviour – *the outcome positively reinforces the action*. Conversely, if the outcome is something unpleasant, the dog is *less likely* to perform the behaviour – the outcome is aversive – it could be said that the dog has experienced *positive punishment*. *Negative punishment* involves the withdrawal or absence of a previously rewarding experience when the action is performed, which will *decrease* the likelihood of the behaviour being performed again. *Negative reinforcement* involves the withdrawal or absence of an aversive experience when the action is performed, resulting in an *increase* in the likelihood of the behaviour being performed.

Ineffective Positive Punishment

On the face of it the solution seems simple – if your dog does something that you don't want it to do, make the outcome aversive and it will not do it again! But behaviour is not that simple! What was the dog's motivation for the behaviour in the first place?

If, for example a dog is behaving aggressively towards a neighbour because it is frightened of him – reacting in an aversive manner, such as the owner shouting, hitting the dog or pulling on its collar, may make the situation worse as this may further increase the dog's arousal. In some instances an owner's response can also act as a reward for the dog by giving it attention. A dog's fear may also be increased because something 'nasty' happens to it when the neighbour is present *(being punished)*! The next time the dog sees the neighbour it may be more frightened and react in a more aggressive way to try and scare off the neighbour . . . it gets reprimanded again . . . its fear further increases . . . and so the spiral continues.

Unfortunately, all the normal and usual human reactions to aggressive dogs tend to increase it. In situations where a dog shows signs of aggression, trying to calm and comfort it reinforces the aggressive reaction and fear if present. Punishment increases the existing fear and thus leads to more aggression (Jones-Baade). However, by employing positive reinforcement methods the dog can be taught that it is rewarded when it does not react to the neighbour and, over time will start to associate the presence of the neighbour with something pleasant happening *(a process called desensitisation and counter-conditioning)*.

Let's look at the common scenario of the dog that has ignored the recall command on a walk – when it eventually returns to the owner it is severely reprimanded. In essence, the dog has just been positively punished for returning to its owner! Dogs make an instant association with a particular action therefore, the dog does not understand that it is being punished for not responding to a command it heard ten minutes ago – it is just aware that it has approached its owner after having a nice time running around, sniffing in the forest, and the owner is immediately 'aggressive' towards it!! Because approaching the owner has been aversive – the dog will then be more wary of returning to the owner next time,. . . which makes the owner more frustrated and cross and results in the dog being told off more vehemently. A stage is eventually reached where the dog is reluctant to come back to the owner at all, so the owner resorts to walking the dog on the lead. If, rather than reacting with positive punishment, the owner had positively reinforced the dog with, say, a food reward when it had eventually returned, the dog would have learnt that coming back to its owner meant something pleasant and would, in future, come back more eagerly when called.

Some owners believe that a dog should be instantly obedient because it respects them and that giving it rewards is bribery. However, bribery is very different from positive reinforcement. Bribery entails the use of a lure to encourage a particular action – positive reinforcement entails the giving of a reward *after* the action has been performed – a subtle but important difference.

Effective Positive Reinforcement

Most people immediately connect positive reinforcement, or reward training, with food. However, dogs are motivated by many things and what is rewarding depends on the drives, needs and temperament of the dog as well as the situation at the time. Obviously, where a dog is highly motivated by food, it makes sense to use food rewards. But if the dog has just eaten its dinner, it will probably be less motivated by food and a toy or verbal praise may be more appropriate. Owners are sometimes concerned that their dog will become overweight if they utilise food rewards. This need not be so – a simple way to overcome this is to measure out the dog's food for the day and use part of the daily ration for positive reinforcement. Many dogs are now fed on a complete dry diet which can easily be carried in pockets – ready to positively reinforce the dog. It is, however, important to have a variety of treats so that if the dog behaves particularly well, a more highly prized reward can be given. That is, there should be *scales of reward*. The better the performance of the required behaviour, the more appetitive the rewards should be. These can vary from pieces of chicken, ham and cheese to raw carrots and home-made liver cake. Touch, eye contact and reacting to the dog, as well as toys etc, can also be added to positively reinforce a particular action.

The breed of dog will play a part in its motivation and it is important for owners to know the origins of the breed as this will affect the dog's behaviour and general character. For example terriers were initially used *(and still are in rural communities)*, for hunting. Most are extremely reactive and like squeaky toys, which they delight in shaking. Border Collies, bred for sheep herding, generally enjoy chasing and running whilst Cavalier King Charles Spaniels, which originated as companion dogs, particularly enjoy physical contact. There are numerous breed societies available to the dog owner and the greater the knowledge of the breed the easier it will be for the owner to find the right motivation. However, there will always be exceptions to the rule!! For example, whilst it is generally accepted that Labrador Retrievers are highly motivated by food – there are some that are not! It is vital that owners are aware of what motivates their dog in order to use the appropriate method of positive reinforcement.

An important factor to consider with positive reinforcement is the type of reinforcement schedule used i.e. how often the dog is rewarded. Whilst Gross *(1992)* defined five types of reinforcement schedule (continuous reinforcement, fixed interval, variable interval, fixed ratio and variable ratio), the two most commonly used in training dogs are continuous reinforcement and variable ratio:

• *Continuous reinforcement* – where each desired response

is reinforced e.g. every time the dog sits, it gets a piece of kibble.

- *Variable Ratio* – reinforcement is given spasmodically, so the reward is unpredictable e.g. the dog is rewarded for sitting once, but the next time it is rewarded after sitting three times, then possibly twice etc. – there is no set pattern to when the reward is given.

As it is important that the dog learns what is required of it, continuous reinforcement is used initially to teach the dog a particular command or action. Once the owner is confident that the dog has learnt what is required and it is responding each time, the rewards can then be given on a variable ratio basis. Variable ratio can be likened to gambling – the dog responds extremely well in the hope that it will be rewarded with a 'win'. If variable ratio coupled with a variance in the value of the reward, *(i.e. sometimes the dog receives a piece of kibble and sometimes it receives a 'jackpot' – a handful of chicken)*, the dog's response will become even more reliable. This is because, sometimes, the dog is being positively reinforced and other times it is being *negatively punished (it is not receiving it's expected reward)* which leads to the dog becoming frustrated, trying harder and subsequently being positively rewarded for the increase in its behaviour. *(This frustration effect is covered in more detail later in the chapter)*

An extremely effective method of positive reinforcement is through the use of *clicker-training* whereby the dog forms an association of reward with the sound of the clicker through classical conditioning. Once the association is formed, this becomes a highly enjoyable and stimulating method of training for the dog as, in essence, it initially has to figure out for itself what it is being rewarded for. With a clicker it is easy to 'click' at the precise moment that the required behaviour is performed. Verbal or other signals can then be added as the dog starts to perform the action and it will make an association between the two. Eventually the dog will learn that performing the action when the signal is given will make the click treat occur but they do not occur if the click treat is not given. Clicker training also helps the owner to perfect their timing because they have to be vigilant and watch their dog in order to 'click' when it performs the required behaviour. There are a number of publications now available on this effective and fun method of positive reinforcement/ negative punishment training.

Inadvertent Positive Reinforcement

Clients often tell me that they don't have time to be constantly rewarding their dog. However, many owners are masters of positive reinforcement – but don't realise it! Imagine the following scenarios with the family dog, Ben:

- Ben wants to play, but his owners are busy getting ready to go out for the night and take no notice of him. Ben happens to spot some training shoes lying on the ground, picks one up and starts to playfully throw it in the air. One of his owners instantly reacts and starts to chase him. Ben finds that this is great fun and runs around the house 'ducking and diving' with the owner chasing him in a frantic manner. The owner chases Ben because he is desperate to get his expensive shoe back. Ben views this interaction as a game and through repetition he could learn that it is a good way of getting attention. *The owner positively reinforced the dog's behaviour of running away with the training shoe.*
- An owner is sitting watching TV and Ben comes approaches and lays his head on their lap. The owner

absent-mindedly strokes Ben's head. Eventually, the owner stops stroking his head and Ben nudges the owner with his nose and they immediately respond by stroking him again. *The owner has positively reinforced the dog's attention seeking behaviour.*

- The owners are eating dinner and Ben approaches the table. One of the owners glances at and makes eye contact with him, which encourages Ben to sit and look 'lovingly' at them. Ben continues to try and catch his owner's attention by placing his head on the owner's knee *(which he knows has worked in the past)* at which point, another member of the family may comment, *'Don't be mean – give Ben a bit of your steak'* to which the owner responds *'I'm not supposed to feed dogs from the table . . . am I Ben?'*, looks at him and slips him a bit of his steak. *Ben has been positively reinforced for begging at the table.*

- The owner returns from work, walks into the house and Ben rushes up and starts to jump up at them in excitement. The owner is delighted by this effusive show of affection and interacts with Ben by greeting and patting him. *The owner has positively reinforced Ben's jumping behaviour.*

All of these scenarios have involved inadvertent positive reinforcement on the part of the owner and made Ben more likely to behave in a particular manner when the situation and similar situations are repeated. Yet, more often than not, when Ben has been lying calmly or chewing a bone *(rather than his owner's training shoes!)*, he has been totally ignored. Many owners tend to interact with their dog when it is, from their perspective, 'misbehaving' or instigating interaction, but ignores them when they are behaving well!

None of the types of behaviour mentioned in the previous scenarios are necessarily a problem and many owners happily accept them. However, imagine a scenario where a young girl comes into the house and Ben bounces up to greet her, as he has done so for many years, and knocks her over. The girl cries, the owners are mortified and immediately punish Ben by shouting 'Bad Boy!' and giving him a smack. Ben is totally confused because he has behaved in his normal way, for which he has always been rewarded, but he is now being told off. Ben is unsure of the situation so he resorts to behaving in a manner that he has previously been rewarded for – he jumps up more vigorously!! The owners, who are still attending to the crying child get cross with Ben and banish him to the kitchen. When Ben is eventually let out, he approaches his owners, cowering, tail between his legs and ears flattened. The owners then 'forgive him' as they believe he is looking 'guilty' because, to them, he looks as if 'he knows that he's done wrong'! Ben *does not know that he has done 'wrong'* – he has been stressed by his owners' behaviour, he anticipates more owner aggression and the resultant emotional state of fear is manifested in his body language. This situation could have been avoided if Ben had been accustomed to being positively reinforced for acting calmly when the family came home rather than for leaping about.

The Dangers of Positive Punishment
So why not use positive punishment? Surely the dog needs to know 'who is boss'? There are an ever increasing number of gadgets and 'quick fix' solutions available for owners with problem dogs and a considerable number of these involve the use of positive punishment, some confrontational *(directly involving the owner)* and some

remote *(the dog does not make a connection of punishment with the owner)*. These range from check chains to collars that automatically squirt citronella spray when a dog barks to collars that deliver an electric shock. All of these methods can be misused either accidentally, or intentionally, and can result in pain or other distress for the dog.

Another possible repercussion of aversive devices can be the development of fearful behaviour towards an unrelated object or environment. For example, if a shock collar is activated to stop a dog barking and at that same instance a child walks past, the dog may associate pain with the presence of that child and start to display fear aggression towards them next time it sees them. I have seen a case where an owner had installed a special fence to stop her boisterous Collie wandering off *(This device works by giving the dog a shock through a special collar when it gets close to a boundary)*. During the consultation it emerged that when the other dog in the house, a timid Collie, had the special collar placed on him he wouldn't even go into the garden! One could argue that the device had worked, but to such an extent that one of the dogs was now frightened of going outside at all.

Whilst accurately executed positive punishment, implemented by someone who is aware of all the possible dangers, can be effective in stopping unwanted behaviour, most people do not have the expertise to implement it without potentially detrimental effects. The *timing of positive punishment is absolutely crucial* and a delay of its use can have profound implications for the effectiveness of the technique (Solomon, Turner and Lessac) and the welfare of the dog. The average dog owner rarely 'punishes' their dog immediately the behaviour is displayed – a prime example being that of the dog that urinates in the house when it is left alone and is then punished some hours later when the owner returns home. The dog will not make any association between the punishment and its previous urination, but it may start to make an association between the owner's aggression and their return home. The problem may well escalate as the dog displays increasingly stressed behaviour manifested in its body language when the owners return. If it finds that this behaviour stops or prevents their aggression it could learn to use it to avert its owner's 'aggressive' behaviour towards it on future occasions. If a dog is extremely stressed a possible consequence is urination and the dog may start to urinate as soon as it sees the owner – which would exacerbate the situation further!

Another point to consider is how severe should the punishment be? A verbal reprimand? The use of aversive tasting substances such as hot mustard or chilli sauce smeared on items to stop a dog chewing? A smack with a rolled-up newspaper? Hitting it with a stick . . . or worse? Positive punishment is intended to be aversive, but it is clearly undesirable to inflict more pain than absolutely necessary. But what is necessary?

If the punishment is too mild, it may be ineffective, and the problem may then not be remedied easily by switching to a stronger punishment. In a study, pigeons were punished with an electric shock for pecking at a key that produced food (Azrin, Holz and Hake). When the shock was set at 80 volts the behaviour was totally suppressed. When the voltage was set at 60 volts, however, it had little effect, and when its intensity was then increased in gradual steps, the pigeons still continued to peck at the key for food even when the voltage had reached 300 volts! If punishment intensity is set at low levels initially and then increased gradually it may prove ineffective because the dog adapts to the gradual increasing intensity. The optimum level of punishment is thus the minimum level that is effective in suppressing the behaviour – but choosing that level is a skill that few people can claim to have. However, even if they have it, is it necessary or desirable? What are the implications for the dog's welfare and its relationship with its owners?

Cognitive and linguistic capacities have been found to play a role in determining human reactions to punishment, but dogs do not have linguistic capacities! A study by Cheyne (1969) found that in a group of children who were given a number of toys and then punished for playing with a particular toy, those children who had an explanation given to them as to *why* they mustn't play with the 'forbidden' toy were far less likely to play with it in a future situation when compared to those children who were just told 'No' and not given an explanation. Sears, Maccoby and Levin (1957) found that mothers who made extensive use of reasoning, reported punishment to be far more effective than those who used punishment alone. Whilst many of us tend to speak to our dogs on a daily basis, and they will respond to certain words such as 'Sit' or 'Down', they do not understand what we are saying. Therefore, in a situation where a dog is being reprimanded, a reasoned explanation of why they are being punished will have no effect whatsoever and the dog will just become confused and may develop emotional responses related to stress and anxiety in other aspects of its relationship. In extreme cases, a dog that feels particularly threatened may resort to defensive aggression towards the owner in an attempt to stop the physical punishment and in anticipation of it. This can then result in an extreme breakdown of the owner-pet relationship, and in some cases, the euthanasia of the dog.

It is important to remember that, as detailed previously, if a dog is displaying fear aggression, positive punishment can lead to an increase in that fear and an increase in the intensity of the behaviour.

If a dog is being consistently positively-punished it can also result in an increase in stress. Animals *(and humans)*, that are placed in a situation where they have to deal with a threatening situation will experience a number of physiological changes, one of which is the increased production of adrenal gland hormones. This physiological response is useful as a short term response in an aversive situation as it results in increased blood flow to the muscles and provides quick mobilization of energy for vigorous movement, but it is harmful in the long term. Prolonged increases in the level of glucocorticoids *(cortisol)* can damage muscle tissue, inhibit growth, inhibit the inflammatory response and suppress the immune system (Carlson). Animal research suggests that stress can encourage the growth of malignant tumours and exacerbate allergies and auto-immune diseases. Equally, prolonged stress can impair the learning process (Dweck & Licht). Therefore, if a dog is continually placed in a situation where it is being positively punished by its owners, its physical as well as its cognitive well-being is compromised.

The Action of Negative Punishment

So far I have talked about positive reinforcement and positive punishment. There are, however, two other learning scenarios to consider – that of *negative reinforcement* and *negative punishment*. Liebermann (1993) defined negative reinforcement as an increase in the probability of a response due to the *removal of an aversive stimulus,* and negative punishment as a decrease in the likelihood of a response due to the *removal of a reinforcing stimulus.*

To illustrate these two scenarios, let us consider the situation of two children playing – a boy and a girl. The girl is very spoilt and used to getting her own way. She starts constantly pulling the boy's hair, telling him that she will not let go until the boy gives her a sweet. Here we have a situation where negative reinforcement is being utilised – the pain (positive punishment) will continue until the boy reacts in the required manner – relinquishing a sweet. This particular method of learning, negative reinforcement, is, obviously, not one to be recommended as its welfare implications are related to positive punishment.

On another occasion the girl continually pesters the boy for a sweet because this has always worked with her parents! Her pleas, cries and tantrums are incessant, but the boy takes no notice, she becomes increasingly frustrated and her behaviour escalates, she stamps, cries and throws herself on the ground but all to no avail and the boy continues to ignore her. Eventually, exhausted, she stops her tantrum and sits down calmly at which point the boy gives her a sweet. Although her tantrums had previously worked, in this instance, she will have started to learn that tantrums will not result in obtaining a sweet (negative punishment), but sitting calmly will (positive reinforcement).

Negative punishment involves what is termed *the frustration effect* where the subject – be it a person or a dog – when faced with a situation where he/she has previously been rewarded for their behaviour, will get increasingly frustrated and will try harder but can become angry or depressed if the expected reward does not materialise. There appears to be a strong emotional component – think of the angry man kicking a vending machine when his expected can of Coke doesn't materialise despite frantically pressing all the buttons or the frustration that you feel when you turn the car ignition on and the engine is dead! You repeatedly turn the key and possibly start shouting at the car before you accept that the engine will not start. Obviously in these two examples, in reality, the behaviour would not be extinguished as the vending machine will eventually be replenished with cans of Coke and the garage will fix your car so that when you turn the key the engine will start again. However, if, for example, the vending machine was never replenished again, the man would eventually learn that kicking or repeatedly putting money in that particular vending machine was ineffective. As he would not receive the 'reward' of the Coke can, he would eventually stop trying – his behaviour would have been extinguished.

Usefulness of Negative Punishment

Unlike positive punishment *(where an action results in an aversive outcome),* negative punishment entails the absence or withdrawal of an expected reward. This method can be utilised to stop unwanted behaviour that has previously been either consciously or inadvertently positively reinforced. Negative punishment can be a difficult concept to grasp, I believe, because the terminology is confusing. The words negative and punishment both conjure up images of something aversive, whereas the concept could be described more simply as non-reward or the ignoring of a behaviour that was previously responded to.

To illustrate this, let us return to our scenario of Ben, who instantly leaps up at everyone who enters the house. How can we stop that behaviour?

In many instances owners will resort to waving their arms about and shouting at the dog to 'Get down' or, possibly, hitting it. The first will inevitably prove to be ineffective, because Ben is stimulated by the increased interaction (positive reinforcement) and will continue to jump up. The second action, hitting the dog, will confuse him and he will re-double his efforts to display the behaviour that he has learnt is rewarded by his owner – jumping up. This scenario continues on a daily basis with the owners becoming ever more frustrated and the dog increasingly confused. So what can be done?

Employing negative punishment will initially result in frustration but eventually lead to extinction of the behaviour. All people entering the house must completely ignore Ben when he leaps up – including no eye contact, ie. they need to behave as if Ben doesn't exist. Initially, Ben will be frustrated as he will not receive the response that he expects and will increase his efforts i.e. the behaviour will get worse. (It is at this stage when I am explaining the procedure to clients that they interject ' But we've tried ignoring him and it doesn't work – he just gets worse!' When I ask how long they persevered with the 'ignoring regime', they will invariably tell me that the dog became such a 'pain' that they couldn't ignore him any more and they gave up.) It is vital that the owners do not succumb to interacting with him as otherwise Ben will learn that if he leaps up for long enough and hard enough, he will eventually get a response. However, by completely ignoring him every time he is excited *no matter how long his excitable behaviour persists for*, and positively reinforcing the calm behaviour by greeting him, Ben's excitable behaviour will eventually be extinguished and calm behaviour, which is differentially reinforced will be more likely to occur.

The problem behaviour will have been solved through the use of negative punishment coupled with positive reinforcement, without the use of any positive punishment and without damaging the relationship between dog and owner. Obviously, if Ben had been positively reinforced for sitting calmly when family or visitors arrived at the house from the outset, the problem behaviour would never have developed in the first place!

Negative punishment techniques can also be combined with *conditioned avoidance*, where the dog learns that it will fail in a particular action if a certain stimulus is present. Conditioned avoidance can involve the use of training discs (five small cymbals on a key ring). Once the discs have been properly introduced, the dog will associate failure *(non reward)* with the sound of the discs and will stop whatever behaviour it is displaying when it hears the sound. For example, a dog jumping up to get food from a kitchen work surface can be made to jump down and ignore the food by the owner chinking the discs or throwing them to the ground. Training discs can be utilised to stop all manner of behaviour from some types of aggression to jumping up.

Difficult Situations

Positive reinforcement (coupled with negative punishment) can be successfully used in most situations to overcome problem behaviours. This can sometimes be difficult to imagine when you have a dog on the end of a lead pulling its owner over in an effort to bite the next door neighbour!

Aggression is one of the most common behaviour problems in dogs and one that owners frequently react to in an equally aggressive manner and positively punishing their dog for. Quite often, if the aggression is directed towards another dog or person outside the home, the punishment given to the dog will be as a result of embarrassment and the owner feels that they have to be seen to tell their dog off for its behaviour. I always advise owners, difficult though it may seem, to concentrate on reacting to their dog in the correct manner, rather than fulfilling another person's expectations. If a dog has behaved unsociably it is always important to apologise calmly and politely, but then concentrate on the dog and try to think why it behaved as it did.

Many owners are reluctant to muzzle their dog as a precaution if it displays aggressive behaviour. I would suggest, once again, this is primarily because they are embarrassed or worried about what other people think or because it does not look very nice! However – if an aggression problem exists, it is far better to ensure the safety of all by placing a basket muzzle on the dog. This will also enable the owner to be more relaxed because they will know that, should their dog behave aggressively, it will not be able to cause any serious injury. It is important, however, to gradually introduce a dog to wearing a muzzle rather than forcing one onto it distressing situations.

Many potentially serious situations can be easily avoided, such as sheep chasing. If an owner knows that this is problem behaviour, the simplest solution is to ensure that the dog is walked well away from any sheep and when necessary, placing it on a lead.

As a general rule I recommend the implementation of the following solutions, dependent on the scenarios involved.
• If the unwanted behaviour is not serious, but something that the owner wishes to modify, such as incessant attention seeking (barking, leaping up etc.), ignoring the behaviour (negative punishment) followed by positive reinforcement of the desired behaviour is

normally effective. However, if the dog has an emotional need it may be necessary to provide an outlet for its attention seeking behaviour, such as the use of an interactive key – where the dog learns that picking up a specific object, such as a particular toy, will result in attention from the owner.

• If behaviour needs to be immediately stopped, the use of interruption, distraction and positive reinforcement can be implemented. For example, let us look at the scenario of a dog that is about to chase a cat (which is a very rewarding experience for the dog!). A distinctive sound, such as a squeaky toy, can be used to obtain the dog's attention, before it starts to chase. If this is immediately followed by encouraging the dog, in an excited manner, to come to the owner and be positively reinforced with the squeaky toy and subsequent game, this will result in the dog concentrating on its owner rather than chasing the cat. Eventually, the dog will find it more rewarding to ignore the cat as it will have learnt that it receives stimulating interaction from its owner.

• Some dogs respond more readily to the use of a '*leave*' command. It must be reintroduced in increasingly stimulating circumstances so that the dog learns that it is positively reinforced for its appropriate reaction to the signal of 'leave'. For example the owner can progress from calling their dog away from a stationary toy, to one that is moving slightly to calling it away from one moving at speed. Eventually, when the dog is placed in a situation, where it may have previously chased a cat, for example, it will return to its owner once the signal 'leave' is given and, due to the implementation of positive reinforcement, its motivation to chase the cat will be reduced.

• Where a behaviour must be stopped so that positive

reinforcement of an appropriate behaviour can occur the use of *conditioned avoidance techniques*, such as training discs, can be extremely effective, as they can help owners deal with more simple behaviour problems. However, it is important that owners properly understand the concept before using them. Children must be prevented from playing with the discs, as indeed they must be prevented from playing with a clicker, as this will confuse the dog and make them ineffective.

Problem behaviours, particularly more serious ones, can be very complex and involve a number of contributory factors such as the history of the dog, the relationship with the owner, how long the behaviour has been displayed, how much stimulation the dog receives and so forth. It is vital to accurately assess the motivation behind the behaviour and this may result in recommending a number of methods of behaviour modification, such as the use of conditioned avoidance, an interactive key, house lines, etc. It is not possible to detail methods for specific problems in this short chapter and those that have been included here are only examples. Owners who are worried about their dog's behaviour should seek professional behavioural help from an APBC member in order that the problem is properly understood and an appropriate behaviour modification plan can be recommended and properly implemented.

Summary

I will not deny that aversive methods can sometimes be effective in stopping a dog's inappropriate behaviour. But in most instances where positive punishment is implemented, it is mis-timed and without awareness of the possible consequences. These may vary from increasing the unwanted behaviour, such as in fear aggression, to the dog forming an association of pain with an unrelated event, such as a child passing by. Frequent positive punishment also leads to stress, both for the owner and the dog, and a breakdown of the owner – dog relationship. Surely it is better to have a dog that responds to signals because it has learnt that compliance results in something pleasant, rather than a dog that responds because it is frightened and wants to avoid punishment?

REINFORCEMENT AND PUNISHMENT	
POSITIVE REINFORCEMENT	**POSITIVE PUNISHMENT**
A rewarding outcome for a particular action: egs. • *When the dog sits calmly, it is stroked.* • *When the dog is not attention seeking it receives attention* **CONTINUOUS REINFORCEMENT LEADING TO A VARIABLE RATIO REINFORCEMENT SCHEDULE** **STRENGTHENS OWNER- DOG RELATIONSHIP**	An aversive outcome for a particular action: eg. • *The dog is jerked on the lead when it does not sit.* **POSSIBLE ASSOCIATION OF PAIN WITH UNRELATED STIMULUS** **TIMING AND INTENSITY OF PUNISHMENT MUST BE ACCURATELY IMPLEMENTED** **WELFARE IMPLICATIONS** **CAN COMPROMISE OWNER- DOG RELATIONSHIP**
NEGATIVE REINFORCEMENT	**NEGATIVE PUNISHMENT**
The absence or withdrawal of an aversive experience when the dog displays the required behaviour egs. • *'jerking' on lead stops when the dog sits.* **COMPROMISES OWNER- DOG RELATIONSHIP** **WELFARE IMPLICATIONS**	*The withdrawal of a rewarding outcome where the action performed was previously positively reinforced: egs.* • *Ignoring the dog when it jumps up.* • *Ignoring the dog when it seeks attention.* **IMPLEMENTED ON A CONTINUOUS NON-REINFORCEMENT SCHEDULE**

Someone once said to me that they couldn't stand by if they came across a group of boys throwing stones at a window, they would have to 'sort them out'. Whilst they did not clarify what methods would be used 'to sort' the boys, I rather suspect they would have included some sort of confrontational positive punishment.

My aim, and that, no doubt, of my APBC colleagues, would be the same – to stop the boys – but our approach would be different. In the first instance, we would interrupt and distract them from the behaviour but then try to find out why they were behaving like this. Boredom? Frustration? They didn't like the person living there? Whatever the reason . . . once ascertained, the core motivation could be assessed and the problem behaviour dealt with to obtain a long term resolution. Admittedly, instant positive punishment may well stop the boys' behaviour on that day – but it would not have dealt with their motivation and if the motivation to perform the behaviour is stronger than the fear of the potential punishment the behaviour will continue. However, once the motivation is known the behaviour can be re-directed into an appropriate, more motivating, positively-reinforced action, thereby resulting in a cessation of the initial unwanted behaviour.

The majority of owners have dogs as pets. They are considered to be a part of the family, trusted and faithful companions. However, all too often, we totally ignore them when they are behaving well, be it lying calmly, interacting well with people and other dogs, or not chasing the cat. But we instantly respond and interact with them when they start to behave inappropriately. Dogs do not speak the same language as us we cannot explain things to them, they will only learn from our interaction with them. So it is important that we, as a supposedly superior species, through patience, understanding and positive reinforcement, nurture, develop and continue to maintain that most cherished and potentially rewarding of relationships with man's best friend.

6

Scent Communication in the Dog

Charlotte Nevison and Rosie Barclay

Are you embarrassed when your dog thrusts its nose into the groin of a visitor to the household? Has your usually obedient canine ever been known to run off apparently oblivious to your calls then stop and roll delightedly in the remains of a long dead creature or fox scat? Do you have to wait because your dog has refused to move until he has checked out and cocked his leg at every lamppost, or seen your otherwise placid pooch turn into a master escape artist when the bitch next door is in heat? Yes? Then you are not alone. To us these behaviours can be inexplicable, confusing, inconvenient and possibly repulsive, but these behaviours indicate that our canine companions are motivated to respond to scents that we are either unaware of, or do not appreciate! In this chapter we will unravel some of the mysteries of the sensory world of dogs by illustrating how their sense of smell differs from ours, why scents are important to them and how misunderstandings about scent relate to some of the common 'problems' we experience with our canine companions.

I Sniff Therefore I Smell . . . The Olfactory Capabilities of Dogs

Before discussing what important information is conveyed in scents we need to think about what scents are and how animals' gain information from them. You will be aware that objects may smell, as may liquids and that some smell more than others. You may also become aware of an odour without deliberately sniffing at something. To this end all animals are basically alike but what is your brain actually responding to?

Scents are airborne molecules (volatiles) that have typically been released from a solid or liquid source. They may consist of a variety of volatiles released at different rates depending on factors such as the weather and whether or not they are being sniffed! When inhaled volatiles are detected by specialised cells which send signals to the brain. These signals may then elicit a behavioural response *if* the volatiles convey information that is relevant to the animal at that particular time.

Although we detect odours in the same way, in comparison with most other mammals humans are pretty poor at detecting scent. It is estimated that dogs have a 100-1000 times greater potential to detect odours than us because they possess more receptors for scent. This allows them to detect volatiles at a lower concentration in the atmosphere; they can more easily detect sources which release volatiles slowly or that are further away.

Making Sense of Scents – What Is the Relevance of Smell to Dogs?

Scientists have demonstrated that dogs and other mammals have a much superior sense of smell to us and are better detectors than any machines ('artificial noses') that have been developed to date. Such research is largely driven by the fact that detecting scents can be of considerable use in human society. Dogs are trained to detect drugs, firearms and distinguish human scents during search and rescue operations and even detect cancer from human urine samples. However, though these skills are remarkable, they are *learned* responses acquired through the process of *training*. They reflect how amazing the canine sense of smell is but they do *not* tell us why it has evolved to be so special. Put simply, dogs are excellent scent detectors because this skill was of use to their wild ancestors.

Although dogs have been domesticated for centuries, scents that were naturally motivating and reinforcing to their ancestors are likely to still influence the untrained behaviour of the domestic dog, though responses may be tempered by breeding, training and whether their needs are fulfilled in the domestic environment.

What Scents are Important to Dogs, and Why?
Two of the key objectives for ancestral dogs was to locate food and defend physical resources as part of a cohesive social group, all within a complex and possibly large area. Scents played a key role in achieving these objectives.

SNIFFING FOR THEIR SUPPER
For a dog 15,000 years ago there were no easy meals and no bins containing tasty scraps! Potential prey would have been scattered and constantly moving over rough terrain, not neatly penned up in fields. In short, the chances of *seeing* potential meals from a distance were slim.

Hearing would have been of limited use because a dog's hearing extends to ultrasonic frequencies, which are easily obstructed by objects including hilly terrain and trees which may have been more important when dogs were evolving.

Fortunately for ancestral dogs (and other predators) potential prey species, or their waste products or scent marks, would have smelt as they released volatiles. Carcasses also smell. Such scents would travel through the atmosphere, around potential obstacles, and would become stronger as they were approached (thus indicating where the source was). Once the food was located the remaining problem to ancestral dogs was acquiring it. The ease at which this was achieved would depend on factors such as its abundance, size and the environment but evidence suggests that much of the time they would have needed a little help from other group members, which in turn required an efficient means of communicating with each other.

THE SOCIAL SIGNIFICANCE OF SCENT

As discussed elsewhere in this book, much of our understanding of ancestral canine behaviour comes from observations of closely related wolf species and modern day feral dog groups. Pair bonds have been reported and are especially important when considering relationships such as the bitch-pup bond. However, free-living dogs typically associate into larger groups, which consist of a number of individuals.

Although members within a group are socially bonded, territories are such that individuals are not constantly in each others presence. It is therefore essential that group members are able to effectively recognize each other as part of their social group upon meeting. It is also important that they are able to communicate with each other remotely. Individuals that do not belong to the group need to be identified, and where necessary excluded, as they may compete for vital resources or mates. Scents play a major role in achieving these objectives.

WHAT INFORMATION DO SCENTS CONVEY AND HOW?

Scientific understanding of scent marking behaviour is largely drawn from studies of rodents, in particular mice, which are easier to study than large groups of dogs. From a combination of body scents and the distribution and properties of scent marks it appears that mice can determine an individual's identity, sex, relatedness, challenge potential, health status and how long it has been since a scent mark depositor was there! They can then use this information to determine how to respond to particular individuals if and when they meet and whether or not it is safe to stay in the area.

It is likely that dogs may also gain social, temporal and maybe even health information from scent marks which, with visual, tactile and auditory support, may determine their behaviour in both domestic and feral situations. Our understanding of the social relevance and behavioural effect of dog scents, though currently not as detailed as it is in mice, is growing. Recent advances, including behavioural and molecular studies of a pheromone secreted by bitches that calms her pups, has given us a more detailed understanding. By creating synthetic versions of these pheromones, such as dog appeasement pheromone, (DAP) we also gain potential tools which, with care and appropriate guidance, may be used to modify behaviour in a beneficial manner in a domestic environment. In the following section we review what we do know about how dogs use scents to communicate and how we can use it to gain a better understanding of our domestic companion and their 'problems'.

Scents and their Sources

Upon meeting dogs use a combination of postures, sounds and body scents to gain information about each other (Who is this? Is he a threat?).

The big advantage of scents over visual and auditory signals is that they can be deposited in the environment as marks which gradually release volatiles to convey social information about the depositor in his or her absence, much as we leave messages with a pencil and paper. Unlike the perception of sound and vision scents are less disturbed by the environment and have a relatively long life.

Odours that provide socially important information in mammalian species are known to be released from many locations on the body including the mouth (saliva), urogenital tract (urine, vaginal fluid) or released from specialised secretory glands near the skin surface (oily or watery cellular secretions). Dogs may use all of these substances to convey social communication, though to human observers the deposition of urine marks is the most apparent marking behaviour and has thus drawn the most attention.

URINE MARKING

Urination by dogs serves two functions, the removal of potentially harmful liquid waste from the body (elimination) and to convey socially significant, individual specific, information to other dogs (marking), typically in the absence of the marking animal.

Males mark in a very characteristic manner by depositing their urine in brief directed bursts by raising one of their hind legs; high if the urine is directed at a vertical target and lower if at a horizontal target. Urine may not be eliminated on up to a third of these occasions (often also observed in pet dogs), though figures vary widely between studies. Occasionally males may squat to urinate, particularly if castrated, which is the typical marking posture of females. Occasionally females may raise the leg into a more male-like posture.

In feral dog colonies, and indeed our pet dog population, adult male dogs urinate about three times as often as females. Most of the time males will expel their urine as marks but females mainly urinate to relieve themselves but may mark to advertise their reproductive status.

The positioning of scent marks can be particularly revealing when it comes to determining their purpose. Detailed studies have shown that marks are deposited mostly at courting sites. The sexual basis of urine marks is supported by studies, which have shown that males are more attracted to urine odours of females than vaginal or anal sac scents, particularly when females are in heat. In the presence of females males will keenly and possessively overmark female urinations probably to convey ownership (though mating in feral groups is typically not exclusively between known breeding animals). However, overmarking (and queuing to do so) is also seen within sexes and is often seen in domestic dogs and wolves. It is not clear whether this serves as an indicator of challenge potential (what the researchers interpreted as status) by the over-marker, or to make a 'family' mark at significant points in the territory.

Marks are also frequently deposited at feeding sites and

territory boundaries with nest sites being rarely marked. This suggests that the function of scent marking is two-fold. The first is to attract mating partners (both sexes) and claim them (males). The second is to claim and defend territorial borders (which in dogs will often partially overlap with other groups) and feeding sites. Nesting sites are probably not marked with urine as to do this would draw attention to youngsters.

Though all individuals do appear to mark, there is variation in the number of times different individuals of the same sex mark. This is likely to be related to the researchers assessment of their position in the social hierarchy (the subject of what constitutes canine hierarchy is discussed elsewhere in this book) being higher for high status individuals (who have priority to mate), in particular the males. Studies have shown that individuals who attain a high-status in the group (particularly males) generally mark at a higher rate, and further increase their marking frequency when they are challenged. It is quite possible that such changes in social behaviour are related to changes in the animal's emotional state. From a functional perspective in natural environments depositing marks at territory borders takes time and energy and exposes individuals to greater levels of stress and danger. Therefore the ability to constantly do this is thought to reveal the individuals superior qualities and thus i) be attractive to the opposite sex and ii) deter or combat challenges (and thus the chance of injuries) from other males. In mice individuals of a lower status will typically inhibit their marking response, depositing their urine less often and in larger amounts which appear to act as a message to higher status individuals that they do not pose a challenge to them. Whilst it is unclear whether this happens in dogs the release of large volumes of urine (and also faeces) is widely interpreted as being an indicator of anxiety in many species which is likely to correlate to an individual's social status.

FAECAL MARKING AND ANAL GLAND SECRETIONS

When a dog defecates a small amount of fluid may also be secreted from the anal glands. Rarely, anal glands may also be voluntarily discharged without defecation, but such behaviour is much more common in the more distantly related cousins of the dog such as hyena. In wolves, deposition of faeces along territory borders and trails suggests that they may play a role in designating ownership of an area.

Both faeces and anal gland secretions have been considered as sources of socially significant scents in dogs relating to sexual attraction, territory marking or signalling of danger. But what is the evidence?

Anal sac secretions contain a solution of lipids, particulate material and bacteria which help to produce an odour. It is thought that the number of volatiles identified in anal sac secretions and the between sample variation is sufficient for unique combinations to occur in different dogs and could therefore be used as a label to convey ownership of territories. Although deposition patterns in wolves suggest this is a possibility in that species there is virtually no evidence for this in feral or domestic dogs. There appears to be little attraction of potential mates to anal gland secretions, at least relative to urine marks. Anecdotal evidence suggests that discharge of anal sacs occurs with fear or anxiety inducing situations. This has not however been supported by examining the responses of dogs to these odours.

111

OTHER POTENTIALLY IMPORTANT SOURCES OF
SIGNIFICANT SCENT

The sources of scents discussed above, and their relevance
to dog communication, have drawn most attention from
scientists interested in theories of animal behaviour. Other
pheromones, such as that secreted by bitches which acts to
pacify their puppies, have raised interest not only because
of what they reveal about canine behaviour *per se*, but also
because of their potential use in an applied context in the
domestic environment. However dogs secrete many other
substances which could also be important to them, some of
which are described below, which have yet to be studied in
detail.

Upon meeting dogs are typically drawn to investigate
the ano-genital region where they may be particularly concerned with information conveyed in urinary odours. They
are also drawn to investigate the facial region though
clearly this involves an element of danger! Licking may
also occur, which in wolves is usually directed by the more
inhibited individual of any pair. Domestic dogs are also
often seen by their owners to lick other areas of each
others bodies (notably mouths and ears) and dogs will
often also lick their owners which may be rewarded or discouraged by owners. Dog saliva is known to have some
anti-bacterial properties and thus licking may have some
health benefits and mutual grooming of each others hard-
to-reach areas may play a role in social bonding. However
saliva, distributed by licking, may also have a communication role. Proteins found in the saliva of dogs have a
notable degree of similarity to those that mice, and other
species, use to communicate their identity. It is therefore
plausible that saliva may be used to convey identity, ownership or group membership.

Domestication and Domestic Environments –
Its Influence on Scent Communication between Dogs

Given the sensory differences that exist between dogs and humans, it is remarkable that by and large we co-exist so amenably. Through the gradual process of association for mutual benefit that lead to the domestication of dogs we have managed to learn to understand each other to some extent, but not entirely.

During the process of domestication of dogs we have altered many factors that influence how dogs behave. Dogs rarely live with groups of other individuals of their species, typically living as the only canine or one of a pair in a household. By not allowing dogs to breed freely we have influenced how dogs behave by generally not select-ing those with unfavourable characteristics to breed from and tending to select for juvenile and/or highly specific characteristics. By using procedures such as neutering we have influenced the hormonal profile of our companion dogs which has many profound effects on behaviour, such as reducing sexual behaviour. Management of dogs as part of our households also removes the requirement for dogs to perform natural behaviours (though not neces-sarily the motivation) and training serves to shape the behaviour of animals through reinforcement. Evidence also suggests that when living with humans dogs often override olfactory cues with information gained visually which indicates a sensory shift on their part to commu-nicate more effectively with their primarily visually orientated owners!

Given all of these influences on the domestic dog, how important is their sense of smell and scent communication to them. How, when and why does it impinge on our gen-erally positive relationships with our companion dogs?

Speaking Different Languages –
When Scent Communication between Dogs and
Humans Breaks Down

We have discussed how it is part of a dogs natural behaviour to mark the territory in which their group resides. However, in a domestic situation there is a fundamental difference – the humans that form part of the 'group' do not urinate to mark the territory borders or to show their challenge potential. Furthermore they do NOT appreciate the canine members of the group performing these tasks at important sites (from the dog's perspective) in the territory, restricting them to specific sites such as the garden! How is this interpreted by the dog? This may depend on the initial human-driven toilet-training process and the relative challenge potential of the dog in the household group. A puppy, who doesn't yet know better or is unable to help him or herself, can perhaps be forgiven for urinating in the house. But when an adult dog, who the owner believes should know better, releases urine in the house conflict occurs between animal and owner.

INAPPROPRIATE ELIMINATION

When an owner asks for advice on inappropriate elimination they are normally referring to a breakdown in toilet training. Their dog has begun to urinate and/or defaecate in areas the owner finds unacceptable. This doesn't mean that it is necessarily showing abnormal behaviour just that the dog is going to the toilet in the wrong place. There are a plethora of differential diagnoses for inappropriate elimination. These include certain medical conditions, age related problems, environmental change, incomplete house training and faulty learning, breed type, inadequate management, substrate preference, over-excitement, fear-related urination, attention seeking behaviour, challenge potential related and anxiety problems. The diagnoses related to scent-marking are challenge potential marking, attention-seeking behaviour and some anxiety problems.

Since inappropriate elimination can be a complicated area to diagnose successfully it is important that the owner seeks professional help in dealing with this problem. The dog should first be examined by a veterinary surgeon to rule out any medical problems. There are many common medical disorders related to urinary problems such as, urinary tract disease, congenital or anatomical malformation, endocrine disorders and neurological abnormalities. Problems relating to inappropriate defaecation include bacterial and viral conditions, obstructions, food allergies, age related problems, activity levels and changes in gastrointestinal motility. If there are no obvious medical reasons the owners can then be referred to a qualified behaviourist.

SCENT-MARKING
Both adult males and females may display this behaviour. To differentiate between scent marking and other diagnoses it is important to look at the amount, frequency, directionality and locality of the urine deposited. When urine and associated scent secretions are used for communication only small amounts are usually deposited frequently and directed at prominent places, though animals may release more in environmentally and socially stressful situations. Often the dog will mark after investigating and sniffing several spots. Commonly reported places are the sides of furniture and walls, standard lamps, upright vacuum cleaners, shopping bags and visitors trouser legs. It is also common for dogs to jump up onto

a bed and urinate or defaecate on the duvet. There are different motivating factors reported for inappropriate scent marking behaviour. Confident dogs will often mark for territory maintenance. Others may have learned that if they perform these behaviours they get attention and some dogs may mark as a consequence of feeling anxious. It is therefore, important that the underlying factors are clearly understood before employing a behavioural modification programme.

CHALLENGE POTENTIAL MARKING AND ATTENTION-SEEKING

Both urination and defecation may be used to scent mark, which is normal canine behaviour, but inappropriate to owners when it occurs within the home. Among females scent marking is affected by their hormonal state and they will pass small amounts of urine more frequently when in pro-oestrus, oestrus and post parturition. Marking behaviour can be reduced by neutering and 'problem' marking can be prevented or at least decreased in over 50% of male dogs by castration. However scent marking is not controlled exclusively by current levels of reproductive hormones circulating in an animal's bloodstream; hormones produced in other parts of the body play a role and social responses and emotional state may also influence physiological processes leading to the exhibition of such behaviour patterns. In multi-dog households and free-ranging groups over-marking is common and more confident females will urinate over urine marks made by males or other females to reaffirm their challenge potential. The dogs that tend to show this confident marking behaviour also tend to show other challenging postures towards other dogs such as direct staring, assertive body language and vocalising. If scent-marking behaviour occurs within a multi-dog household it is important to address this problem by reassessing the relationship between the dogs and treating them accordingly. However, challenge potential marking may also occur in a home where only one dog resides if there is social stress. Commonly a dog will mark on walls or furniture and even a trouser leg in front of their owners. The leg raising posture on its own may also as act as a visual threat display. The dog may also show other responses to social stress such as aggressive behaviour or persistent attention-seeking.

Since the smell of urine in a particular place may encourage the dog to over-mark, it is advisable to clean the areas used thoroughly. Biological washing powder is often used and the area wiped with surgical spirit. There are also several commercial furniture deodoriser sprays which clients have recommend.

CASE STUDY
A relevant case study involved a Jack Russell dog called Bob who lived with a recently married couple who, on demand, gave him every thing he asked. They constantly played throw the squeaky toy for him, gave into his food begging, allowed him to persistently climb all over them and at night he would artfully push the husband out of the bed. As soon as either of them stopped interacting with him he would run into the hallway and defaecate on the mat. This made both owners react and although Bob was told off he still gained a reaction from them albeit a negative one which maintained this behaviour. He would also walk boldly over towards any visitors, cock his leg up and urinate all over their trousers. It was this particular behaviour that the owners initially asked advice about.

The owners were tactfully asked to ignore attention-seeking behaviour and, if necessary, gently push Bob away without looking or talking to him when he tried to initiate any form of unwanted contact. They were then asked to solicit all interactions with Bob by first asking him to come to them, to sit down and to wait. He was not allowed to have a fuss, a treat, his dinner, lead on or a toy until told. Bob soon learned that to get nice things he had to sit down calmly and wait first. When visitors arrived they were asked to ignore him and for a few weeks he was kept on a lead in this context. All marks were cleaned using biological washing powders and a deodorising spay and he has not marked since.

ANXIETY PROBLEMS
Some dogs, typically of low confidence and/or a nervous disposition, may mark as a consequence of feeling anxious or unsure of a new environment. There are many reported cases of dogs that urinate or defaecate when they first enter new surroundings. Once they have performed the deed they rarely repeat it. To discourage this behaviour a deposit of their smell can be laid prior to the dog entering by wiping an old towel/blanket that the dog has been sleeping on around the walls and furniture. It would also be advisable to have the dog with you and supervised until it has gained in confidence.

Some dogs find it hard to cope when left and may show anxiety problems such as increased vocalisation, destructive behaviour and elimination problems. Usually these involve full bladder elimination of urine and not scent marking as such. However, it has been suggested that some dogs scent mark to increase their feeling of security. Therefore, the separation-anxiety problem needs to be addressed before the marking will decrease. Not all problems are mutually exclusive and can figure a multitude of underlying causes and it is important that each dog should have a medical check up before referred to a qualified behaviourist. In some cases appropriate drug therapy may be recommended by the veterinary surgeon. In these cases and concurrent to the behavioural modification programme the dog appeasement pheromone (DAP) may be of some use.

ROLLING IN SCATS

Rolling is a normal canine scavenging and hunting and possibly confidence related behaviour, which is generally ignored unless the odour is unpleasant to humans – such as rotting fish! It is fair to say that a dog's motivation to pursue such odours is usually very high so if the owner does not wish their dog to smell like a rotting fish they should anticipate rolling behaviour and work further on reinforcing effective leave and recall behaviours. After a bath rolling is likely to aid manipulation of the dog's scent glands and spread their scent over their body once more. To stop them rolling in the mud directly after a bath it is advisable to keep them indoors and give an alternative low-odour substrate to roll in, such as their bedding.

ROAMING

Intact male dogs roam more frequently than castrated ones and will sniff out sexually receptive females over long distances. The obvious dangers include being injured or killed on the roads and causing third party injuries or fatalities. Unsupervised dogs also pose a health hazard and fighting can also break out with other roaming males. It is therefore, important that this problem is addressed. Since roaming behaviour is usually hormonally modulated castration will help to decrease this behaviour by about 90%. However, this behaviour also has a learned component, and the longer the dog has been showing this behaviour the less likely castration will help in reducing it.

If after neutering the dog is still an ingenious escapologist the owner should make sure it is sufficiently exercised and its attention needs are met and that it wants to stay with the owners rather than being made to. Eight feet high fencing sunk into concrete and surrounded by razor wire may physically stop the dog from running away but it will not stop the dog wanting to. A qualified behaviourist will be able to help with this problem.

STEALING SCENT IMPREGNATED ITEMS

Many dogs pinch scent impregnated items, such as tissues. This can include the owner's underwear which the dog may refuse to give back. Why a dog would choose these items over and above others we can only surmise. However, these items of clothing do carry a relatively large amount of the owner's scent. They are also a handy size to fit comfortably into a dog's mouth. These items when taken, procure an explosive reaction from the embarrassed owner. Therefore, the dog quickly learns to make the association between the item and the reaction of their owner, which may be interpreted as play, such as tug-of-war. The initial motivation to take these items thus changes from a natural interest in group odours to a learned association, which gains the reward of attention. One solution to overcome such an attention seeking motivation is to dissociate the item with the reward. Instead of becoming all flustered and chasing the dog around the house simply ignore the dog and fetch something 'better' (determined by your

Photo: David Appleby

1. The Golden Retriever – selective breeding has enhanced its retrieving instinct

Photo: David Appleby

2. Exposure to a wide range of stimuli in early life enables the puppy to develop its maintenance set

3. Puppies must learn good manners, when greeting each other on or off the lead

4. Playing games with a dog is a great way of increasing the bond and helping children and dogs understand each other

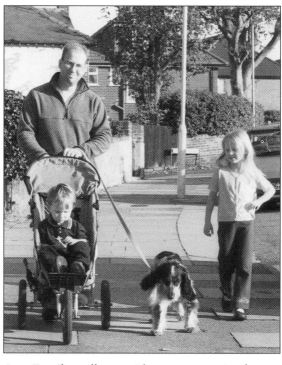

5. Family walks provide an opportunity for dogs to form positive associations with children

6. It is important to reinforce rather than ignore convenient behaviour

7.   Clicker training provides an effective means of reinforcing the
      required behaviour

8.   Scent communication allows signalling in the absence of the
      signaller

excitement and interest) such as the dog's favourite toy or treat. Become very interested in this 'first prize' and the dog will quickly lose interest in it's now 'second prize' and will come trotting over. Do not try and force the item away from the dog, simply ignore and wait until the item is dropped. Carefully pick up the soggy underwear and sometimes ask the dog to sit, wait and take what you have. The dog will learn that to receive nice things it has to sit and wait first and that pinching items of underwear does not achieve the desired result of attention. Some dogs will take items impregnated with the owners scent to lie on, or lie on those that are already available at times of stress, such as during separation anxiety. This acts as a maintenance stimulus that increases their capacity to cope.

Conclusion

Scents are important to dogs, much more so than to us. They convey messages relevant to their (ancestral) survival and social behaviour. Such messages are not obvious to us humans, as we do not have the same ability or motivation to respond to them. Unfortunately this can and does sometimes lead to a breakdown in dog-human communication. Scientific studies of the behaviour and physiology of dogs and the experience of behavioural counsellors in deciphering how to deal with dog-human misunderstandings has helped to unravel some of the secrets of the canine world of scents. However there is still much to learn which requires us to keep our eyes, ears and most pertinently our noses open to what our dogs are telling us.

7

The Relationship Between Emotions and Canine Behaviour Problems

David Appleby and Jolande Pluijmakers

To explain the development of behaviour problems and know how they can be prevented and treated it is necessary to understand how they are influenced by emotions. Emotions are experienced as either positive or negative. They result from a match or mismatch between events in the environment and the animal's interests. A mismatch gives rise to a negative emotional state, such as fear or anxiety, which causes the individual to act in order to cope with a threat to its physical or emotional homeostasis. The emotions experienced are a physical reaction in the form of an increase in the parasympathetic (relaxation) or sympathetic (arousal) part of the autonomic nervous system and corresponding decline in the other, as discussed in Chapter 2. If the increase is in the sympathetic the state of arousal provides a fast mobilization of energy. Hormone release strengthens the autonomic response by extra increasing the blood flow and making more glucose available to the muscles. The behavioural response consists of the muscular movements, postural changes, facial expressions and sounds appropriate to the situation that elicits the emotion.

The function of emotions is to increase an individual's chance of survival. This is most clearly illustrated by the autonomic and hormonal reactions that prepare the body for actions such as flight or fight when faced with a threatening stimulus. For our and the dog's pre-mammalian ancestors this was all emotions consisted of, however other responses and functions evolved. Expression of emotional states occurs through behaviour, body language and sounds, which serves a useful social function that has survival value because they may influence the behaviour of other individuals due to it being easier for them to judge what is the individual feeling the emotion likely to do. For example, a threatening growl and body language indicative of arousal directed at an individual that tries to take a valuable resource may cause it to withdraw. By communicating emotional states the stability of social groups is enhanced decreasing the potential for conflict and increasing the chance of survival. An example of the latter occurs when distress vocalisation, triggered because a stimulus or stimuli the individual is dependent upon is not available, normally results in reunion with the other group members.

Emotions are subjective, which means that although the process has the same characteristics for all members of a species, it does not result in the same outcome for each individual. How an animal is affected by a situation is influenced by many factors, such as genetic and environmental influences on the development of the brain, past learning experiences and the complex information processing procedure within the brain. The latter is influenced by how different parts of the brain respond to different stimuli and contexts. From this it follows that the coping response, through which an individual minimizes an aversive effect of a stimulus or context, may differ enormously.

Genetics

Recent research of the heritability of personality in identical twins shows that personalities are established by both genetics and environmental factors. The sensitive period of behavioural organization, discussed in Chapter 2, is an example of how genetically predetermined biological development combined with environmental factors, such as learning experience, have a profound influence on the emotional development of an individual.

That an emotionality can be inherited has been shown experimentally in the parallel breeding of two lines of Pointers. Dogs from Line A were confident in the presence of people and remained so. Dogs from Line E were shy of people. Puppies that were bred from dogs in this line showed severe aversion of human company. Cross-bred descendants (one parent from Line E and one from A) were initially confident in the presence of people but subsequent generations of this cross resembled line E and were shy of people. This experiment suggests that in these lines the inheritability of nervousness was dominant. This is important because the genetic potential for nervousness and other emotional states helps to determine how an animal will react in a situation.

By observing the emotional expression of his children and comparing them with people living in various isolated cultures around the world Darwin (1872/1965) was able to conclude that there is an innate species-typical repertoire of emotional expression. Although all members of a species inherit the full repertoire of behaviour from their ancestors, in dogs, as discussed in the chapter about their ethology, the capacity to perform particular tasks has been increased in some and allowed to diminish in others through selective breeding. Therefore some behaviour is easier to trigger in some breeds than in others.

The variability of behaviour from breed to breed was shown in research conducted in the 1960s when a number of breeds were reared in identical circumstances and tested at different ages for their behavioural performance. Fox Terriers, for example, were described as aggressive and non-fearful, Basenjis as aggressive and fearful and Beagles as non-aggressive and non-fearful. If breeds are genetically predisposed to display patters of behaviour, it follows that the type of problem behaviour displayed by dogs will often be a result of their genetic predisposition. For example, a Border Collie has such a strong drive to herd that it will not only herd sheep but any livestock whose movement stimulates the response. When given inappropriate learning opportunities, it may also be stimulated to chase cars, cyclists and joggers.

The genetic predisposition for the performance of behaviours that are suggested to have no obvious function and can be triggered by stress provides a further illustration. For example, a Doberman may suck its flanks whereas a German Shepherd Dog may chase its tail.

The chemical processes associated with the emotional state of the bitch may affect the whole litter. For example, a bitch that is stressed during pregnancy is more likely to give birth to emotionally reactive puppies. Within a litter individual differences are further refined. There are differences in the some types of behaviour and behaviour problems displayed by males and females. It is suggested that within gender differences in behavioural potential may occur due to positioning in the pregnant bitch. For example, a bitch that lies between two male puppies during pregnancy may have some masculine characteristics when compared to a bitch that lay between two bitches.

Learning

A few stimuli, typically those that present a significant, life-changing situation, automatically produce an emotional reaction. Other stimuli do not cause a spontaneous emotional response, instead it has to be learned through classical conditioning during which the emotion becomes associated with a neutral stimulus. Once the learning has taken place, which means that the brain structures have integrated the emotional state of the animal with cue- and context-related information. If the stimulus is associated with an evasive experience the individual will become frightened when it encounters the conditioned stimulus or a comparable situation again. As every individual strives to maintain a state of emotional homeostasis, they will display behaviour to try to avoid or remove this negative emotion. When they have learned an effective coping strategy that minimizes the aversive effect or avoids the stimulus the emotional response will no longer occur.

OPPONENT PROCESSES

Emotional states are automatically opposed by mechanisms of the central nervous system, which reduce the intensity of both pleasant and unpleasant feelings. Following the introduction of something pleasurable a positive emotional response begins and quickly rises to a peak (state A). It then slowly declines to a level where it remains if the quality and intensity of the stimulus is maintained. When the stimulus is removed the reaction gives way to a qualitatively different type of reaction (state B), which reaches its own peak of intensity and then slowly disappears with time and goes back to normal. These reactions are opponent so if A is pleasant, B is aversive; conversely if A is aversive B will be pleasant. The B-process will acquire more strength if frequently elicited. It will also rise to a higher peak and take longer to decay than the A process. In addition the effect of the A process decreases through repeated stimulation.

The effects of frequent state A stimulation may be at work in dogs displaying attention seeking behaviours such as stealing. Giving a lot of attention to a dog pleasure (state A) is created. Repeated stimulation of state A results in an increasing craving for attention when it is not available (state B) and more attention is required to return the dog to state A and so on. The craving can become so strong that the dog may learn the strategy of stealing objects, such as the remote control for the TV, purse, the owner's glasses etc. when the owner is pre-occupied because it finds that they will always respond when it picks them up.

How emotions are processed

Emotions are commonly experienced when a stimulus, context or behaviour activates memories. These memories differ from factual memories because they contain response information and activate the associated systems resulting in an experience of the emotion. What takes place between the perception of a stimulus and the response to it can be described as a data processing within the brain. This processes is assumed to be present because the response to a stimulus varies.

The process involves a number of stages, which are summarized below:

127

1. Assessment of the stimulus: eg. is it familiar or unfamiliar? how do the properties of stimulus compare with other stimuli?
2. Assessment of context: The context is evaluated to assess the potential to remain in control of emotional homeostasis in the presence of the stimulus/event.
3. Evaluation: Based on information from previous stages the importance of the stimulus/event is calculated, which determines the priority it is given. If it is high ongoing behaviour will be disrupted.
4. Behaviour plan: An action plan is generated based on information from previous stages.
5. Physiological changes and selection of behaviour.
6. The behaviour is performed.

In contrast to what this might suggest the process is not an isolated, linear process. The recording of information and the evaluation of its relevance are continuous. As long as an organism is awake it will scan/explore its environment and compare the stimuli it encounters with its interests, which are not static.

It is normal for an emotion to result from the process described above. However parts of the process can be 'left out' or overtaken by other processes and they can be interrupted at any time. For example, context evaluation may be skipped in response to sudden, intense or unexpected events in which there is no time to evaluate. Everything between context evaluation and overt behaviour may be left out if the individual has learned a response that is strongly established or innate response.

Anxiety, fear and phobia

Anxiety, fear and phobia result from a loss of emotional homeostasis causing activity in the sympathetic autonomic nervous system. Variation in behavioural signs is dependent upon the characteristics of the individual, the situation and the intensity of the emotion experienced. Some dogs become restless, pace, tremble, pant or salivate when anxious. Others may bark, whine or seek comfort from maintenance stimuli e.g. by trying to hide where it feels secure or seeking proximity to the owner. When fearful they might freeze or try to escape by flight e.g. digging into locations, scratching at doors or windows. If necessary a dog may display defensive behaviour in an attempt to escape from the situation and return to emotional homeostasis.

Not being able to predict the outcome of a situation is frightening. Inability to predict makes it difficult or impossible for an animal to organise its behaviour so that it can respond appropriately. As a consequence of a sense of loss of control its emotional homeostasis may be disrupted resulting in it becoming anxious or frustrated, which can lead to the development of behaviour problems ranging from major to what might be considered to be nuisance behaviours, such as whining or jumping up. However, apparently mundane problems caused by an animal not being able to predict its environment can develop into something more serious.

A dog's need to be able to predict its environment and have ways of controlling how it is affected emotionally also applies to the social group; in most cases this will be the family and any other pets it lives with. If owners are inconsistent in the way they interact with their dogs, at worst appearing to swing from a 'Jeckel' to 'Hyde' personality because they don't understand their pet's perception of their behaviour, the inability to predict its owners and know how to respond may result in it becoming insecure, the potential consequences of which include defensiveness.

This is especially true if it is subject to higher levels of stress through reduced ability to cope because poor behavioural organisation and/or members of the family are inclined to be punitive or heavy handed.

Here is an example of what has been discussed above. If a dog is never fed tit-bits from the owner's plates it will not expect anything and will not look for it. However, if someone in the family sometimes feeds the dog from their plate they create an expectation that it will happen again and so the dog learns to sit and watch people when they are eating just in case. If it takes longer than he expects for someone to feed him, he might try to remind them that he expects to get something by pushing under the person's arm. If they do not react immediately the frustration he experiences will make him try harder and push more or try another strategy, such as barking. Not such a problem you might think but the dog's ability to predict its environment and know how to respond will take a turn for the worse if he then finds that one or more family members, perhaps not those that did the original feeding, tell it off for 'begging'. The extent to which it is affected emotionally will depend upon the degree of threat the dog perceives itself to be under and its confidence in its capacity to cope. It may try to cope by avoiding its owners in this context, giving them what they intended, or by becoming defensive. Any telling off for being defensive that follows will only confirm to the dog that it was right to fear its owners. In response to the owners becoming unpredictable the dog may become defensive on other occasions, perhaps, initially in the context of the owners being in possession of food it is interested in and they tell it to do something but this can generalise to when the dog perceives itself to be confronted in other situations, just in case the owners turn nasty.

We have already seen that an animal can inherit a pre-disposition to fearfulness. We have also seen that early experiences during the sensitive period of behavioural organization and the formation and development of their maintenance set have a profound effect on the animal's capacity to maintain in emotional homeostasis in a changing environment and when it encounters novel stimuli.

Learning experiences through classical and instrumental conditioning are relevant to the development of anxiety, fear and phobia, its generalization to other situations and any increase or decrease in the intensity of the animal's emotional response. For example, a dog that is always friendly to other dogs may, as a result of being bitten by a large black dog, start to avoid large black dogs. This behaviour can easily generalize to all types of dogs through emotional responses being triggered by dogs that look almost the same size and colour, then ones that are less similar and so on. If the dog has several more unpleasant experiences it may become phobic of other dogs and display an intense and immediate emotional reaction every time it sees another dog. Conversely if the subsequent experiences with other dogs are positive the same dog may learn to overcome its fear.

Some stimuli have a higher genetic potential for triggering an immediate and intense fear response. The large number of dogs that suffer from noise phobia suggests that an example of such a stimulus in the dog could be the fear of sudden loud noises such as fireworks or gunshots. When this stimulus is encountered there is often an immediate and intense emotional reaction before the brain has identified what the noise is, where it is coming from, if it indeed is threatening to the individual and what the best behavioural strategy would be in this situation. All these steps in the emotional information processing procedure are initially skipped because in a life-threatening situation there is no time for extensive and relatively slow evaluation. A fast and immediate reaction may substantially increase the chances for survival. Stimuli present at the time of the bang are easily associated with the negative experience and can subsequently trigger fear because of that association. If, in addition to this, the animal is able to anticipate something frightening may happen because of previous experience, it may become anxious, especially if it does not have an effective means of coping.

Here is an example of how this can influence an animal's behaviour. Some dogs become reluctant to go out to relieve themselves before the owners go to bed or go for a walk after dark because they are concerned that fireworks may re-occur. Except for certain times a year, we can predict that it is unlikely that fireworks will occur but for the dog darkness allows it to anticipate there might be and this association will only extinguish over a period of time during which darkness is not followed by the onset of fireworks. However, the next time they occur at night the dog's capacity to predict its environment and maintain control of its emotional state is further reduced and further still if they occur during daylight. Remaining indoors is no guarantee that the dog will be able to satisfactorily control its exposure to the bangs and emotional state because to a lesser extent they continue to occur. For some this reduction is sufficient to enable them to cope. Others have to use strategies to control their sense of security, such as trying to get close to their owners or hiding in den-like locations. When the owners are not at home fear of the noises or the expectation that they may come can result in separation problems, such as digging into places to hide or escape by any means. If none of these works the dog experiences complete loss of control they may loose toilet control and exhibit other symptoms of anxiety and fear.

The analyses of the context through information processing in the brain can lead to emotions of anxiety or fear. Fear or anxiety results from a loss of control that is experienced when an animal loses too many stimuli from its maintenance set or the changes in its environment make it more reliant on its maintenance set.

For example a dog that has grown up with an older dog and has learnt to depend on it as a means of maintaining a sense of security on walks may behave differently if it has to be walked on its own. It may stay much closer to the handler, check more often if they are still there, scan the environment more intensely and startle more easily. If in addition to being out 'alone' without its giant maintenance stimulus, the other dog, is taken to a strange environment where the dog's capacity to predict and control is further impaired, it may become stressed and start to pant, pace or tremble. Minor stimuli it would normally not react to may cause an intense emotional response, which may lead to it becoming extremely reactive and bark, growl or try to hide behind the owner's legs. It is for this reason that when introducing new stimuli or environments the composition of the maintenance set should be evaluated carefully to prevent it from developing negative associations with the stimulus or the environment whilst at the same time prevent it from being too dependent on any one stimulus.

Separation Problems

Problems that occur when a dog is left at home alone constitute a significant portion of the behavioural specialist's caseload. Until the 1990's the term 'separation anxiety' was commonly used for all separation related problems. However there are many causes for separation problems that are not related to anxiety. One type of separation problem namely 'separation anxiety', has been defined more specifically and is often described as problematic behaviour motivated by anxiety that occurs exclusively in the owners' absence or virtual absence. Rather than being exclusively caused by being left unattended by the owner as suggested by some authors separation anxiety can be defined as apprehension due to removal of significant social or non-social maintenance stimuli which can, for example, be persons, other animal(s) or familiar surroundings.

Fear and anxiety and the opponent process theory play a considerable role in separation anxiety. Commonly reported symptoms such as destruction, vocalization and house soiling are indicative of a loss of emotional homeostasis and some symptoms seem to be associated with attempts to regain access to the lost maintenance stimulus (see table). The opponent process affects the dog's emotional homeostasis because comfort due to social contact (A process) is replaced by negative emotions due to loss of sense of security and control when the dog is left alone (B process). Repetition of these contrasting emotional states results in attention seeking and neediness when the owners are at home and as a consequence a really emotional state occurs when the dog is left alone.

Based on the latter of the definitions of separation anxiety given above, the composition of maintenance sets can be divided into three groups, A, B & C.

Dogs in Group A are those that have remained dependent upon the mother figure or a substitute for them, such as the first owner. This typically occurs where, as a puppy, the dog was not able to explore and develop some independence. This may be caused if the owner is inclined to encourage close contact rather than independent behaviour, where the puppy has been unwell etc.

Dogs in Group B are those that have developed a broader maintenance set of social and non-social things through learning but their emotional homeostasis can still be upset if one or more things they heavily depend upon are removed from the maintenance set and/or their confidence to cope is reduced by a perceived or actual threat to its well-being.

Dogs in Group C have such a broad maintenance set that they can cope with the removal of stimuli from it because enough remain to enable it to stay in a positive emotional state. However they can still be made fearful by the presence of aversive stimuli or other experience. The more complete its maintenance set the better able it is to cope.

Dogs can move between groups B & C. For example, a dog may have a broad maintenance set and not be dependent upon an owner but if an owner stays at home for a long period of time during redundancy or maternity leave their availability, the quality of the relationship and the resultant emotional association may result in them becoming more important in the maintenance set. A dog belonging to Group C may move to Group B after moving house with his owners. Having lost a lot of its non-social maintenance stimuli and being surrounded by unfamiliar stimuli can create a temporally inadequate maintenance set leading to separation anxiety. With time, however, the dog will probably rebuild its maintenance set making it possible to cope with the removal of maintenance stimuli again, and move back to Group C.

The symptoms seen in dogs from these three groups are likely to vary, as shown in the table below.

Group	Time of Onset	When owner present	Behaviour during departing and greeting	When owner absent
A	From Puppyhood. Symptoms start within minutes of being left often immediately after the departure or absence of the social stimulus eg. a person the dog is specifically dependent upon.	Organisation of all activity around one social stimulus. Follows about the home. Needs physical contact, eg. wants to be next to a specific owner. High need for attention/ affection. Tends to stay close to the social stimulus rather than explore on walks.	Distressed when separation is anticipated, eg. trembling, shaking, depressed. May attempt to prevent departure. Over excitement when the owner returns.	Destruction orientated towards doors and windows that give access to the direction by which the social stimulus left. Destruction of items impregnated with the owner's scent, eg. bedding, remote controls, papers, shoes. Vocalisation consistent with distress/relocation.
B	Starts after removal of one important stimulus, several less important stimuli from the maintenance set and/or because the dog has an increased need for the maintenance set eg. due to a new fear eg. rehoming, moving house, left in other room. Only when the dog is left in circumstances where its maintenance set is inadequate.	Dependence on the presence of one or more social stimuli and/or non-social stimuli eg. location in the home. Dependence on social stimuli can increase if separation unpredictable. May show dependent behaviour if there is an increased need, eg. if bangs occur.	Distress during departure and excessive greeting. May attempt to prevent departure. Departure may result in agitation or depression.	If dependent on social stimuli destruction occurs during attempts to gain access to them. If the dog is sufficiently dependent upon inanimate stimuli, it may dig so as to hide or get into rooms shut off from them, such as the owner's bedroom. If fearful they may escape, by any door or window. Vocal due to distress/relocation but may not occur if caused by fear or if the dog is not over-dependent upon absent social stimuli. Defecation and urination may suggest a sense of loss of control.
C	Problems start after an experience that causes fear or phobia that may be associated with the absence of the owners and triggered by it.	No dependency upon specific stimuli. May be fearful of the stimulus when owner is present. The extent to which this occurs may be less because the maintenance set is more complete, behaviour is more ordered and the dog has a greater sense of control.	Distress signs can develop if owner absence is associated with stimuli that cause fear.	Defecation and urination can occur due to emotional state associated with loss of control. Dog may try to cope in a disarranged way or an organised way by trying to escape or hide.

There are other reasons why separation problems occur when dogs are left unattended, the symptoms of which are the consequence of an emotional state. Some problems occur because the dog is not able to anticipate that they are expected to stay, resulting in feelings of frustration and annoyance when they can't leave with the owner. This may result in problems such as barking in a manner consistent with these emotions or giving the cushions a good shake. It is easier for a dog to know that it is going to stay behind and thus avoid these problems if its owners wear different clothes for walking it and going out without it. Even a pair of Wellingtons or a hat only worn when it is going to be taken out will help. The dog's lead being left undisturbed as the owner gets ready to leave and being left behind once they have gone should cause the dog to expect to stay at home. However these clues are missing if the owner hangs the lead with the family's coats so that it rattles every time someone gets a coat or keeps the lead where the dog cannot see it. It is preferable for the lead to hung on its own hook in an area of the house where the dog can always see it and can learn that that if it remains undisturbed it is going to stay behind.

Aggression

Aggression is almost certainly the most acute and threatening behaviour problem for the dog-human relationship. Aggression may be directed towards other animals belonging to the same or another species, familiar and unfamiliar people or objects, such as cars, and serves a variety of functions in an animal's life. In its natural setting aggression has value for the survival of the individual and species. However in the domestic dog the behaviour is often misinterpreted because judgment based on its observation without taking into account how the behaviour developed and is influenced by factors and processes such as genetics, emotional development, and learning experiences. For example, a dog overtly barking and lunging at another dog may be labelled as 'dominant' by the casual and ill informed observer. However, the cause of the behaviour could be fear. Although dogs may initially try to avoid or escape from what they are frightened of they can develop a coping strategy of defensive behaviour to keep it at a distance. The contrast between the emotions caused by what they perceive as aversive (A process) and the sense of relief when it goes away (B process) results in the reinforcement of the defensive behaviour and it becomes increasingly overt. As a consequence of the apparent success, most stimuli go away of their own volition, the dog's confidence is increased and this reduces its fear in that context. This can result in body language that does not suggest fear leading to a wrong assumption about its motivation. A more accurate approach is to describe what we can observe of the animal's behaviour. For example, instead of saying a dog displayed a particular type of aggression it would be more accurate to describe what it actually did, e.g. growled, barked, bit, approached or withdrew etc. We can separately consider its emotional state from one situation to another through analysis of the available history.

Most of the available literature tends to categorise aggression according to perceived motivation, examples include, fear, dominance, redirected, possessive, protective and predatory. Looking at aggression from a biological point of view, this extensive and confusing list of can be reduced to resource-related aggression, fear aggression and aggression resulting from a physical cause.

Social animals spend a vast amount of time together and conflicts may be inevitable because resources are commonly limited and not every animal can exploit them equally. They must communicate with each other to gain access to and utilization of them. To reduce the possibility of fighting species-specific genetically programmed ritualised signals are displayed and increase an animal's and the groups' chance of survival. Vocalisations, facial expressions and body postures communicate the emotions of animals involved in the conflict. Various factors influence the potential for aggressive behaviour, these include, temperament, learning from previous encounters, observational learning e.g. from the mother, hormones, age, sex, size, distance and degree of social inhibition.

Aggressive behaviour caused by competition for resources and directed towards pack members in the natural environment or familiar people or dogs living in the same household, has, until recently, been called dominance related aggression. Descriptions of this behaviour include growling or biting when competing for or protecting resources such as a bone, a favourite place on the sofa or food bowl.

In simple terms a sense of being able to control resources can be expressed as an equation of *Resource Holding Potential* (RHP) + *Value* over *Cost*. This comes down to; do you normally get your own way plus something you want minus any concerns you may have about the ability of an individual that wants the same thing to stop you and cause you harm. If you think you are stronger than the other individual, really want the resource and do not think the other individual will harm you it is likely that you will not feel inhibited by them and will control the resource. If two individuals keep having incidents where this equation comes into effect they should reach the point where you can anticipate who will win and in what contexts this is likely to occur resulting in a ranking system that avoids challenge and conflict.

Ranking systems have evolved in pack society because fighting decreases when everybody knows where they stand. Problems tend to occur in the human/dog relationship when owners convey signals that unwittingly cause their dog to develop a sense of control over resources and become uninhibited in their relationship with them, increasing the potential for agonistic behaviour, of which competitive aggression is one aspect. For example, an owner who readily hands over the food they are eating when their dog demands it may be developing in it a sense of control over the resource of food in that context. Of course this act on its own is unlikely to upset the dog's perception of the appropriate ranking order but if it perceives itself as being able to control resources in a number of situations it may become uninhibited about challenging its owners in other situations. The importance of control is shown in some research in which it was found that dogs are more likely to develop lack of inhibition in homes where their owners acquiesce to their demands (O'Farrell 1995).

Contrary to popular belief fear and anxiety may play an important role in status related aggression. For example, fear of loss of control of those resources that have become important for the dog's emotional homeostasis. Behaviour modification programmes introduced to cure a hierarchy problem, which often involve the owners denying access to resources the dog cares about, can result in depression and withdrawn behaviour. It is thought that this is a reaction to failing to get anticipated pleasures such as portions of the owner's food or to lie on the sofa. However, once the dog has adjusted to the new situation and its expectations are lowered this tendency stops. Nonetheless, it is necessary to reduce expectations so that aggression does not occur as a consequence of frustration when the dog's expectations of being able to have what it want when it wants it are not met. To prevent a negative emotional state of depression occurring owners can request their dog to earn everything it wants or is given by responding to signals to sit, lie down etc. before the resource is given. Thus effectively maintaining the owner's control of the resource.

It has also been argued that many instances of aggression towards people in the dog's social group, perhaps the majority, are not a status problem but related to avoiding an aversive outcome during an associated negative emotion. During social conflict dogs that have become uninhibited may learn to use defensive coping strategies to avoid an aversive event more readily than inhibited (submissive) dogs, which are more likely to choose to flight or freeze as a way of coping.

Anecdotal evidence suggests that dogs presented for the treatment of status related aggression are often found to be nervous and constantly seeking attention from their owners and are described as being generally fearful. When making detailed analyses of the bite incidents, the display of aggression can often be traced to a stressful situation, such as the sound of a loud noise occurring just before the incident. Anxiety may be caused by other chronic environmental stress or a lack of predictability in social relationships or physical environment. If owners add to the stress of fearful or anxious dogs through apparent confrontation, even if not intended, their behaviour may be somewhat disorganised, resulting in a range of possible behaviours that includes excitability, failure to respond to their wishes and defensive aggression. These may be misinterpreted as 'dominant' behaviour.

Physical causes for aggression include, physical and perceptual ability, pain, endocrine disorders, liver or kidney disorders, brain disease and the influence of diet. The strong link between medical conditions and some behaviour problems explains why the APBC strongly believes that best practice requires that medical causes for behaviour problems are excluded or identified and treated before behaviour therapy is introduced. Failure to do so could be to the detriment of the welfare of the animals in our care and therefore all behaviour problems are seen on veterinary referral.

There are two types of behaviour, redirected and predatory, that are often categorised as aggression that should probably not be. Redirected occurs when a dog is in an aroused state, which may or may not involve aggression, but cannot get to what has caused its arousal and it redirects its arousal as aggression, typically onto what has distracted or frustrated it. Predatory behaviour is not an aggressive act but associated with obtaining food, and different areas of the brain become active.

Conclusion
Behaviour is the means by which an animal maintains emotional homeostasis. From the outcomes of its behaviour it learns to display the most effective response for that and similar situations to achieve this goal. Many behaviour problems, including minor ones, are the consequence of disruption of emotional homeostasis and the performance of behaviour that has become disorganized and ineffective

as a coping mechanism or is perceived to be effective but inappropriate for our requirements as pet owners and society in general. Therefore, it is essential that the emotional states of an animal are properly understood and what is learnt included in our approach to preventing and treating behaviour problems.

8
Our Relationship with Dogs
David Ryan

It is a curious aspect of the scientific process that under-standing remains fluid and information is only accepted until it is supplanted by more rigorous, more thoroughly researched, better founded, knowledge. Often the per-ceived wisdom of the day is no more than a working hypothesis, albeit based on the best interpretation of the information to hand.

Throughout the history of scientific endeavour theories are supplanted by other theories as they are subjected to the scrutiny of peer review. Acting on the most up to date interpretations of companion animal behaviour is extremely important to APBC members because as leaders in the field we bear the responsibility for translating best academic theory into best practice.

One of the most significant changes in attitude in rela-tion to the understanding of canine behaviour is the move away from what has become known as the 'dominance theory' of the domestic dog and the punitive training methods that go with it.

In highlighting the move in theory, I make no criticism of the original proponents or followers of it, I was one myself and we were following the best available information at the time, but many leading animal behaviourists are concerned that the dominance model of pet dog behaviour continues to survive, despite the accumulating evidence that it is at best unhelpful and at worst highly detrimental.

The Book of Matthew 13.57 quotes Jesus as saying, 'A prophet is not without honour, save in his own country, and in his own house' meaning that it is very often the strictest adherents of a particular mode who fail to see its drawbacks. Perhaps they are too close to it, are too used to living by it, or have vested interests in continuing to advocate it. Perhaps there is an element of cognitive dissonance in abandoning theories upon which you have staked your professional reputation. However, once you understand that theories aren't forever, critically examining them and changing your mind completely or reinterpreting the facts on which the original theories were based becomes not only inevitable, but actually necessary for advancement. John Maynard Keynes, when challenged on a change of stance, famously said, 'When the facts change, I change my mind. What do you do, sir?' (Malabre, 1995). Everyone should reserve the right to change their mind as they become aware of new facts.

It is an exciting time to be involved with the understanding of domestic dog behaviour, as the late 20th and early 21st century has seen a marked acceleration in the study and application of training and behaviour modification techniques but this means that there is an onus on practitioners to keep up with their professional development or risk being left behind. We should not rely on the remarkable adaptability of the domestic dog to cope with all manner of environments to allow us to continue to with a suspect formulaic approach because it works with some dogs; we have a responsibility to develop methods that will work not just for all dogs, but for all people who own, live with and train them too.

In ethology, 'dominance' is correctly used to describe unequal relationships between individuals where one is able to secure access to resources such as mating opportunities, food or territory. In a competitive interaction between two individuals one will eventually defer, conferring dominance on the other. Repeated encounters that follow the same pattern establish the terms of the relationship, and very often deferential, appeasing or threat-reducing behaviours develop in order to avoid potentially costly conflict. Having established their relationship it is maintained by the submissive individual continuing to defer. Failure to defer as expected sparks conflict and the dominance relationship breaks down into a competitive one.

Dominance is useful in this context because it allows us to predict the behaviour within the relationship with reasonable accuracy. What is important to emphasise is that dominance is not a quality of an individual, but an emergent product of a relationship. Individuals cannot be dominant unless it is conferred upon them by others.

So, if this concept of dominance within a relationship is useful, why can't it be applied to relationships between dogs and owners? The answer is that it can, so long as it is applied in the ethological sense as a product of a relationship, not the quality of the dog.

The problem arises when dogs are judged to be dominant when the meaning is that they are trying to secure dominance over their owner, as though dominance itself was the goal. The inference being that either your dog has to be dominated, or it will dominate you; that dogs are programmed to fight to be 'top dog' or 'pack leader' and you must subjugate them through pinning them to the floor, holding their jaws closed or displacing them from favoured sleeping places and making them watch you eat before they do.

The popularity of the idea amongst dog trainers and owners alike revolves around its simplicity. Dogs are genetically identical to wolves; wolves pack and fight for dominance within that pack; therefore dogs will fight for dominance within their human pack. It's neat and it's catchy, it just doesn't stand up to scrutiny. It supplanted the previous theory because there wasn't one beyond, 'Me human, you dog; me say, you do.' I liked it at the time I was training police dogs because it gave me a reference point beyond the push-pull-and-shout methods. What I didn't like was that it often didn't work. It originated from pulling together a number of sources of information.

- Emerging DNA studies showed that the ancestor of the domestic dog was the wolf – and they still are the same species.
- Studies of dogs showed that their communication by body posture and expression was very similar, if not identical, to that of wolves.
- Observations of captive wolves suggested a pack hierarchy in which the 'Alpha' male and female dominated the others.
- This hierarchy was maintained through constant conflict, often bloody, between members of the pack as they fought for resources.

Adding all that together gave us, 'Dogs behave like wolves and will fight for dominance'. It seemed sensible, but was in fact a huge leap of faith based on misplaced understanding.

Dogs are indeed the same species as wolves, but same

species doesn't mean same behaviour. The (now extinct) aurochs was the ancestor and same species as domestic cattle, but you'd have been seriously ill-advised to try to milk one. Domestication changes not only the appearance, but also the behaviour of animals, as they adapt to fit into a different environmental niche. Nobody doubts the change in appearance from undomesticated wolf to domesticated dog, but recent studies have highlighted that the changes in behaviour are equally apparent once they are examined.

Dogs evolved about 15,000 years ago to fit into the new environmental niche provided by the construction of permanent human settlements. Fixed towns and villages demand a waste dump, which can be a source of nutrition for local wolves. They would not be discouraged by the humans, because they could also be a source of nourishment. Wolves that were able to remain at the dump longer were able to feed and therefore benefit more, setting up a selection pressure for reduced flight distance. Reduced flight distance, as shown in experiments on silver foxes originally by Dmitri Belyaev and more recently by Lyudmila Trut (1999), brings with it all the attributes of the domestic dog, from coat variation, curly tail, dropped ear carriage, a tendency to bark, smaller brains, shorter faces and biannual seasons through to increased levels of the neurotransmitter serotonin, which modulates anxiety and impulsiveness. In short, the domestic dog is a neotonic version of the wolf-type ancestor, a specialised variant that retains juvenile characteristics into adulthood, that evolved into a newly formed environmental niche to scavenge the domestic waste of human settlements.

These adaptations removed the need to operate as a true wolf pack and consequently there is little collaboration in

hunting or in care of offspring, but much more cooperation with strangers, dog or human. Although, given
relevant circumstances, dogs may congregate in groups
around resources, they do not form packs in the cohesive
family way that wolves still do.

The unit of a wolf pack is the family. The 'Alpha' male
and female are the breeding pair, or 'mum and dad', whilst
the rest of the pack usually comprises of their offspring in
varying stages of growing up. They act in concert together,
cooperatively hunting with each other, or staying behind
to look after pups. Proceeds of the hunt are brought back
for the baby-sitters and pups, carried in the stomach and
regurgitated.

The unit of the domestic dog is one dog: one individual
that looks after itself. There is little cooperation in hunting for food, as food is doled out by humans, either
dumped as waste or spooned out of a can. There is no uni-
versal communal care for the pups, either through
babysitting or from regurgitation by other family members.
Indeed, it is an exceptional pup that knows its father,
domestic bitches being promiscuous enough to mate with
many dogs and males being sexually active throughout the
year (as opposed to wolves whose testes and sexual
appetite shrink out of season). The genetic benefit has
swung from heavy investment by the male wolf in pups he
can be fairly sure carry his genes, to no investment from
the male dog that relies on the weight of numbers of his
promiscuous lifestyle to perpetuate his line.

Studies of domestic dogs living together indicate no
linear hierarchy, but rather a more complex interaction
between individuals, where each will secure resources
dependent upon their own preferences. At the time these
studies were taking place the study of wolves continued,

only this time in the wild. It now seems that the previous studies, based on captive wolves, did not correspond to wild wolf behaviour. Captive wolves were often brought from different families and corralled in an enclosure where they could not avoid each other. Resources were dictated by their human keepers and stress levels were consequently extremely high; inevitably competition and conflict were rife. L. David Mech's 2008 article *Whatever Happened to the Term Alpha Wolf?* illustrates that a more rigorous examination of wild living wolves revealed that their social behaviour was built around cohesion and co-operation, not conflict. The model of the wolf's supposed fight for dominance and alpha status was replaced with one where parents and older siblings guide and lead younger off-spring in order to enhance overall genetic fitness.

Recognising that the flawed dominance model arose to fulfil a need – the need to explain why our dogs sometimes come into conflict with us – we cannot supplant it unless we have another workable theory. Fortunately we have.

The domestic dog evolved from wolf behaviour into low level scavenging around human settlements. Whether this was self domestication as a by-product of being in the right place at the right time, artificial selection by humans as a food source or, most likely, a combination of the two, might be an interesting speculation, but still results in the same behaviour. Domestic dogs have plasticity of pheno-type (physical appearance) and also of behaviour, a tendency to friendliness towards, or at least tolerance of, strangers, either canine or human, and low levels of hunt-ing behaviour, because these would be selected against in a human environment (the humans would not tolerate them). The basic domestic dog is a relatively innocuous hanger-on for scraps who owns nothing worth defending.

We don't own this type of dog, however, we own pedigree breeds, or at least in the western world we own dogs that are identifiable as being either a breed or a mix of breeds.

A breed is a type of dog that is identifiable as the off-spring of two parents who were also the same breed, and breeds happen for a purpose. The behavioural plasticity of the original dogs meant that it was relatively easy in husbandry terms to introduce breed tendencies and we now have dogs for all sorts of purposes: sight hounds, scenting dogs, shepherd dogs, guard dogs, earth dogs and lap dogs. Each of these dogs has needs that are based on the function for which they were originally shaped and it is very often their desire to perform that function that brings them into conflict with us. Place a breed designed to be active in a modern family that thinks two twenty-minute walks a day is good exercise, or place a breed designed to sit on the lap in a family that goes out to work for ten hours each day and something has to give. What we are witnessing in 'dominant' dogs is natural behaviour, sometimes competitive in nature, that has been modified through learning. The result is the dog trying to secure something they know is going to have a positive emotional benefit – either to obtain a reward or to avoid something unpleasant. How we deal with those emotions determines our relationship with our dogs.

Like all animals, dog behaviour is often related to their inherent boldness or shyness. In shy individuals behaviour that does not meet owners' expectations is likely to be tinged with fear and in bold individuals the behaviour is likely to be joyously unrestrained. Most dogs' behaviour will be a complex mixture of these two extremes. That complexity is increased because our pets no longer continue to live in their original state as peripheral scavengers but have been refined through selective breeding for specific purposes. By enhancing traits present in the original stock, humans have created dogs whose emotional balance depends on being able to fulfil their desire to exhibit these inherited predispositions, at least to some degree. Although the working traits of these types are reduced during 'pet-ification' – the breeding of more amenable individuals that are more suited to life as a pet (witness the current 'pet-ification' of the Border Collie from a working animal) – the breeding stock continues to throw up specimens in which the original working temperament is strongly represented. This may be a predisposition to chase moving objects, to nip heels, to use aggression to solve conflict, to hold something in the mouth, or any other breed predisposition. The need to perform these behaviours, and their dissatisfaction when they are unable to do so, can steer pet dogs into conflict with their owners.

Family life can also be remarkably inconsistent for a pet, and dogs may focus their efforts on resources that are extremely important to them, but not necessarily to the owner. Lack of consistency proves to the dog that they are capable of deciding the outcome of many, albeit small, interactions. Add in the effects of either a bold or shy character, and other inherited predispositions that need to be satisfied, and you have a dog that can be extremely resistant to their owner's efforts to control their behaviour.

If the perceived solution is to out-dominate the dog and bend them to your will and involves things like rolling them over and holding them down, shaking them by the scruff and competing with them for access to resources, both dog-welfare and the dog-owner relationship are compromised. In dogs where lack of compliance is motivated by frustration at being unable to fulfil inherited needs, where the motivation is fear, such as when the dog has developed a fear of being left by the owner or fear of the owner themselves because of the mistaken application of inappropriate techniques, applying misguided ideas of dominance will increase that frustration and fear, and with it the probable use of aggression. Less confrontationally, standing in the dog's bed to show them who is in charge similarly fails to address the underlying emotional issues.

So what can we do with these dogs that have been labelled 'dominant'? We can regard them as part of the family. Dogs have evolved to live with us, and our basic unit is the family, but as is the case with all family members they need to know what their role is – their position within that family group. All members of our families need to know what the rules and parameters of accepted behaviour are. As parents or guardians it is our responsibility to guide them through that, whilst taking their needs into account. It is no different for dogs – firstly recognise the needs of your own pet and, having taken them into account, work within them to balance their emotions.

The way to do that has already been admirably explained in other sections of this book, so it is unnecessary for me to repeat the benefits of positive reinforcement here, although I would like to highlight my concerns about being labelled a 'positive trainer'. The term is sometimes pointed at me as an accusation of weak, fluffy liberalness by some and sometimes used by others to demonstrate their caring attitude but both these perceptions display a lack of understanding of learning theory. Life is full of little rewards and little punishments; they happen to us all the time and our dogs are no different. It is possible to reward every action but it is also possible not to reward it. Not supplying an expected reward can be so punishing that it can lead to frustration, anger and aggression. It needs to be used carefully.

Every time we prevent our dog doing something that it wanted to do we inflict a little punishment. Standing still to teach 'no pulling on the lead' is a punishment. Painting taste deterrent on the furniture legs so they taste bad is a punishment. Saying 'No' when my toast drops on the floor to stop her eating it is a punishment. Sending him out of the room for jumping at visitors is a punishment.

Punishment is not only acceptable in training dogs, it is inevitable but not all types of punishment are equal and each person has to make their own judgment on what they are prepared to inflict. We should, however, draw the line at punishments that might cause the breakdown of a relationship or have welfare concerns for the dog because a relationship based on fear and competition is at odds with the companion dog ethos and the efficiency of our partnership with working dogs and those used for competition.

What we need is a relationship based on mutual rules. No one gets everything they want as and when they want it. No one gets their way every time. Teaching a dog that they can earn rewards, be they treats, the chance for a game or our affection, for doing the things we want, has the consequence that when they do things we don't want or don't respond as requested these opportunities are withheld because rewards and punishments go together. Rules are necessary, consistency is vital, steadfastness can be appropriate but brutality is not.

Taking on the role of parent or guardian for our dogs means that we can care for them and help them understand how to fit into our families. Living with our pets is not just about training them to sit when told, although that might be important too, it is about consistently guiding their behaviour through the use of many little interactions throughout each day. Being your dog's guardian stops the uninhibited ones getting into trouble, helps reduce stress in the anxious ones and caters for all wonderful temperament types in between. Guiding rather than domineering fosters a mutual respect as you both become aware of each other's needs and work together to fulfil them.

9
Feline Behaviour Problems –
The influence of natural
behaviour
Sarah Heath

Introduction

In a domestic environment many of the natural feline behaviour systems are compromised and cats find themselves living in groups of unrelated individuals, being made to share important resources and denied the opportunity to hide or retreat from situations of potential conflict. In addition their human companions place social demands on them which are at odds with their natural behaviour and often fail to provide for many of their basic instincts. Such constraints on normal behaviour result in stress and tension in our feline companions and, in many incidents of reported behaviour problems, an understanding of feline ethology not only helps to explain how and why the more common problems develop but also offers practical methods for dealing with them.

Feline society

Although the traditional image of the cat as a solitary creature is not entirely accurate it is important to remember that much of feline behaviour is based on individual survival and many fundamental behaviours, such as feeding, hunting, resting and eliminating, are performed in a solitary context and have no social significance. Feline society is based on co-operative groups of females who are related to one another and live together in a mutually beneficial environment, which supports the successful rearing of kittens. Males are usually excluded from these social groupings and live their lives as solitary individuals who only venture into the main social context at times of breeding.

Hostility

Levels of hostility within the related groups are low but intrusion by unrelated individuals poses a potential risk to offspring, increases the demand on essential resources and is poorly tolerated, so aggression to strangers can be intense. However, physical aggression carries with it the risk of injury to all parties and in a society where ultimate survival is an individual responsibility it makes sense to avoid situations, which could result in a decreased ability to take care of oneself. For this reason overt aggression is minimised by the use of elaborate distance maintaining behaviours which are designed to keep strangers at bay and discourage social interaction with individuals outside of the social group. These signals include postural and vocal communication, marking behaviours, such as urine spraying, and elaborate use of eye contact and facial communication, including ear positions.

Affiliative behaviours

Within a social context the reliable identification of fellow members of the social group is vitally important and affiliative behaviours, such as mutual grooming and mutual rubbing, are used to cement relationships and to exchange scent signals. Indeed the mixing of scents results in the formation of a group social odour which reassures individuals that the social group is stable and enables them to relax in close proximity to one another.

The challenge of multi-cat households

Over recent years the cat has increased significantly in popularity and at the same time the number of multi-cat households has also increased. Some of these households are made up of sibling pairs, mothers and offspring and other combinations of related individuals but in many cases there is no such basis to the feline community and cats are being expected to live with total feline strangers. Owners often acquire the cats as companions for one another and when problems of inter-cat aggression begin to surface within the household they are genuinely dismayed and perplexed by the situation. However, natural feline behaviour provides no basis for toleration between unrelated felines and while the high proportion of neutering in the domestic population undoubtedly reduces hostility between strangers, and enhances the chances of feline integration, it by no means guarantees it.

In some cases the aggression between housemates may be manifested as outright physical confrontation and the cats may be taken to the local veterinary practice with torn ears and puncture wounds to prove it, but feline tension can also result in more subtle signs of unease and can contribute to other behavioural issues such as indoor marking, over grooming and even alterations in appetite. It can also be a significant factor in the onset of medical conditions, such as feline idiopathic cystitis, feline orofacial pain syndrome and a range of dermatological and gastrointestinal disorders.

Dealing with aggression between cats in the household
One of the most important steps in dealing with problems of aggression within a feline household is to ensure that each individual has ready access to all of the important resources such as food, water and resting places. Just because cats live in the same human household does not mean that they belong to the same social group and it is not uncommon for multi-cat households to be divided into smaller subunits of individuals, which do not communicate across those social divides. If this is the case it will be necessary to ensure that there are enough feeding stations, watering holes and latrines for each of the social groups to exist as an individual unit and attempts to make these groups share these essential commodities will invariably result in social tension, if not overt aggression. It is also important to remember that cats are solitary feeders and even when relationships are good between cats in the household it is ideal for them to be given separate feeding stations and given the ability to control their own access to food so that they can snack as and when they need to.

Increasing available space
When physical space within the home is limited the amount of available territory can be increased by paying attention to vertical space, and the installation of resting platforms and provision of bedding on tops of wardrobes and kitchen cupboards, can significantly increase the space available. Such structures will also increase the opportunity for cats to retreat and hide when they perceive themselves to be in danger and such provision for the expression of natural coping behaviours will assist in reducing stress and increasing feelings of home security. In addition cats that

are provided with such escape routes will be less inclined to use self appeasement behaviours such as over grooming or excessive eating to deal with their stress and thus passive symptoms of inter cat tension will also be reduced.

Broken relationships

When cats have previously co-existed in harmony but have started to display hostility toward each other it is important to determine the reason for the breakdown in the relationship. For some it may be the result of disruption of their social grouping by the entry of another feline into the home while for others a breakdown in communication may follow a brief period of separation, for example when one of the group is taken to the veterinary practice for treatment.

Regardless of whether the cats have previously co-existed or are being introduced for the first time integration is unlikely to be successful from a position of hostility and a period of total physical separation can be useful in order to allow scent integration through the mixing of bedding and toys. In this way the cats can come into contact with each other without any direct form of threat and once they can accept the scent of their housemate without any signs of hostility the owner can then begin the controlled process of introducing the cats in a face to face situation. While food may be used to induce a positive emotional reaction in the cats in order to facilitate this process it is not beneficial to feed them in proximity to each other and the practice of feeding cats at decreasing distances from one another is not recommended. Indeed the reintroduction process needs to be a passive one in which the cats learn to tolerate the presence of one another rather than engage in direct interactions or form close bonds.

Take your time

Patience will be the main requirement at this stage and any temptation to rush the reintroduction will need to be resisted. During the introduction process the overriding aim is to improve the cat's perception of security within the home and the use of pheromone therapy has been shown to be very beneficial in this context. Cats use facial scent deposits to mark their territory and to surround themselves with reassuring signals and synthetic analogues of these signals (marketed as the product Feliway®) are now available which can assist in relaxing feline inhabitants and signalling to them that their home is safe and secure.

Pressures from neighbourhood cats

Multi-cat households and their associated feline issues are not the only result of increasing feline popularity. Social tension within local neighbourhoods is also heightened by the increasing population density in urban areas. In feral situations the size of feline territories and the density of feline inhabitants will be dictated by the availability of vital resources such as food and shelter, but in domestic situations each cat has both of these resources provided for it by its owner and therefore relatively small geographical areas can sustain relatively high numbers of cats without obvious problems. However, such populations are often very unstable and any minor challenge to their stability, such as the introduction of a newcomer, can result in significant levels of inter cat tension and an increase in fear related behaviour problems in many of the individuals.

Dealing with tension within neighbourhoods

In the majority of cases of tension between cats in the neighbourhood, and especially those that involve overt aggression, it is the owner of the victim that first seeks professional advice. Without co-operation from both sets of owners the prognosis for reform is often guarded. Where aggression primarily occurs outside in the gardens and beyond, management will be aimed at decreasing unwanted encounters between the cats and effective time share systems, where owners allow their cats access to outdoors at different times, can work well.

Increasing the feline appeal of each cat's home is also a useful technique and ensuring that all of the cat's behavioural needs are catered for within the house and garden can decrease their desire to venture further afield and so decrease the incidence of interaction with neighbouring individuals. Improving visual security in the outdoor environment by using potted plants and garden furniture to provide cover for cats as they move about their territory is also worth considering. In some cases the victims are not venturing far from home and the aggressors are actively pursuing them, sometimes even into their own houses. These situations are far more difficult to deal with and when cats proactively attempt to increase the size of their territory it is important to ensure that vital resources are not being compromised in some way in their home, for example by competition from housemates or by a basic lack of provision for their behavioural needs. Once again co-operation between owners will be the key to success and this is often difficult to achieve.

Human demands on feline companions
Increasing feline populations both inside and outside the home undoubtedly influence the incidence and nature of behaviour problems in the domestic cat population but another very important factor to consider is the effect of

changing human expectations on the behaviour of this ideal modern pet.

One of the reasons for the cat's popularity is its relative independence and its ability to cater for its own needs, but while most owners are happy to know that their cat is getting on with its life while they are at work they also expect their pet to provide companionship and social interaction for them when they get home. This desire to engage in low frequency but high intensity interaction with cats raises very specific challenges since the normal pattern of cat to cat interaction is one of very frequent but very low key communication. In other words cats will frequently pass the time of day with each other but do not engage in periods of intense communication, whilst owners spend many hours ignoring their cats but want to cuddle and pet them intensely when they return home. Such a fundamental difference in approach to social interaction is bound to bring tension and education of cat owners in the unavoidable restrictions of natural cat behaviour is essential.

Teaching cats about social contact
It is possible to educate cats to appreciate and value human interaction and the process of socialisation is important in preparing kittens for life in a human environment. This education needs to start early and kittens between two and seven weeks of age will benefit from specific learning opportunities designed to increase their tolerance of intense physical contact from people. Increasing feline acceptance of being picked up and cuddled requires specific preparation since these human actions deny the cat the opportunity to use its species specific defence strategy of flight to protect itself from perceived harm. In order to maximise the benefits of the cat-owner relationship the

manner in which young kittens are handled is critical. Remembering normal feline behaviour it is not difficult to understand why the acts of being picked up and restrained are potentially threatening and periods of lifting, gently restraining and touching the kitten all over its body should therefore be routinely incorporated into any programme of early handling for kittens. Even when this has been done effectively it is important to respect the cat's natural behaviour and keep highly restrictive handling to a minimum, learning instead to respond to feline greeting behaviour and to use vocal interaction to enhance the relationship. Whenever cats do need to be handled the principle of minimal restraint should always be applied.

Pressures of a domestic environment

In addition to the effects that other cats and people have on the behaviour of the domestic cats there are also specific constraints of the domestic environment which lead to further compromise of their natural behaviours and the successful integration of kittens into an average human household is something that requires a certain amount of preparation.

While genetic influences will help to determine how an individual reacts to novelty and challenge in its adult environment, adequate exposure to a wide range of stimuli during the early process of behavioural organisation will also be crucial in ensuring that the cat has a broad frame of reference with which to compare its later experiences. Kittens which benefit from an inherited boldness, a varied early environment and ongoing mental and physical stimulation in adulthood will undoubtedly benefit most from the diversity of the domestic environment and will be less likely to exhibit behaviours which compromise their life as a human companion.

Modifying the human environment to cater for feline needs

In order to minimise the risks of feline behaviour problems it is essential to pay attention to the interactions between cats and between cats and people, but it is also important to modify the domestic environment in ways which will enhance its ability to cater for fundamental feline needs. Studying the nature of feline territories helps to highlight the qualities that cats value and illustrate the importance of things like privacy, choice and hygiene in a feline context.

Providing three dimensional space

The need for access to vertical space not only applies in multi-cat scenarios but also in single cat households and provision of elevated hiding places is important in achieving the feline aim of minimising fear and anxiety within the home. Failure to cater for the natural feline defence strategies of flight and hiding can result in cats feeling threatened and increase the risk of developing problems related to aggressive behaviour. It can also result in chronic feline stress which is manifested in inappropriate self-appeasement behaviours, such as over grooming and overeating, and is now recognised as a significant risk factor for a number of medical conditions. Simple alterations to the home in terms of providing radiator cradles and pieces of bedding on tops of wardrobes and kitchen cupboards can make a considerable difference to the quality of the environment from a feline perspective and in situations where anxiety is causing obvious behavioural changes the addition of a pheromone diffuser to the home will also be beneficial (Feliway®).

Changing to natural feeding patterns

The provision of immediate and unhindered access to

resources is an important feature of feline existence and is equally important in both single and multiple feline households.

In order to achieve this in relation to food it is advisable to adopt a policy of ad libitum feeding rather than stick to a rigid system of two meals a day. This also makes more sense in relation to natural feline feeding routines and to feline anatomy and physiology. Hunting behaviour is a time consuming activity in the wild and cats undertake between 100 and 150 hunting attacks per day over a period of six to eight hours. These attacks lead to relatively small meals and interestingly the success rate of feline hunting behaviour is surprisingly low, with only 10% of them actually resulting in the acquisition of consumable prey. Consequently the ratio of energy expenditure to energy consumption is high and this leads to a very effective level of natural weight control. In contrast, in the domestic environment food is provided without any need for feline effort and the ratio of energy input and output is significantly affected leading to problems of unsuccessful weight control and the risk of obesity. The tendency to provide meals on a twice daily basis can result in cats failing to eat all of their food in one sitting because the capacity of their relatively small digestive tract is exceeded. After all an animal which is designed to eat between fifteen and twenty small snacks over a twenty-four hour period is likely to struggle to consume fifty percent of its daily food requirement in one go. Such "fussy" behaviour is readily interpreted by caring owners as a sign that the cat is not happy with the food and selection of more palatable and more energy dense diets is often the result. Not only does this increase the tendency for cats to hold out for a more palatable option but it also further upsets the balance between energy input and output and runs a very real risk of inducing problems of feline obesity.

Increasing the effort in acquiring food

In order to deal with problems of feline weight control attention needs to be given to both sides of the input : output equation and alterations in the methods of feeding need to be combined with changes in the provision of both mental and physical exercise for cats. The use of multiple feeding stations around the home will encourage the cat to search for food and if relatively obscure locations can be added from time to time this will further increase the challenge.

So called puzzle feeders can be used to increase the effort that is needed to gain access to the food source and such devices can significantly alter the amount of time and effort that the cat has to spend in the process of consuming its daily diet. Commercial puzzle feeders are available in some pet shops but it is easy for owners to make their own from a small soft plastic drinks bottle, in which they make small holes, just big enough to release the pieces of dry cat food, which are placed inside. As the cat knocks the bottle around the floor it will be rewarded with the arrival of the pieces of food and the fact that the bottle is transparent and makes a noise as it moves will help to keep the cat's interest. However, cats do need to learn to use these devices and in the early stages the holes in the bottle should be so large that the food falls out with minimal feline effort! It is also important to ensure that a nail file or similar device has been used to make the internal rims of the holes as smooth as possible before cats are given these feeders to use.

Ensuring adequate water intake

It is natural for humans to consider eating and drinking as

behaviours which happen in conjunction with one another and therefore to place the food and water bowls for their feline companions side by side. In fact cats are averse to drinking water that is found close to a food source and this behaviour stems from the need in the wild to avoid water courses which could be contaminated with the urine, faeces, blood and guts of prey animals. It is therefore important to provide water in separate locations from feeding stations and to ensure that the water is also provided in a form that is attractive to cats. There is very little motivation for the behaviour of drinking in cats, since they naturally eat prey which is rich in water content and therefore supplies the majority of their water need. If water is provided in bowls or in locations which make the act of drinking difficult they will often walk away and there are a range of medical conditions which can be exacerbated by low water intake so this can be very significant. Providing water in shallow but wide bowls which are made of glass, metal or ceramic and ensuring that the bowls are always filled to the brim will help to ensure that there is no taste taint and that the cat can easily see the meniscus of the water as it laps, all features which help to maximise feline drinking behaviour.

Providing adequate mental and physical exercise

Ensuring that cats expend sufficient mental and physical energy during the day is not only important in terms of weight control but also in terms of maintaining physical and mental fitness. Cats are designed to engage in short bursts of high energy consuming activity, which is often related to predatory behaviour, and intersperse these with significant periods of rest and relaxation. Provision of outlets for this energy rich activity is therefore an important part of cat ownership and an understanding that predatory behaviour is stimulated in isolation of the sensation of hunger helps to explain the very real need that cats have for periods of predatory motivated play.

The need to daily exercise

The fact that predation is most likely to occur at times of dawn and dusk means that much of a cat's high intensity activity will naturally occur at these times and failure to make specific provision for predatory responses during the day can result in cats whose night time and early morning activity becomes seriously disruptive to the domestic environment. In addition a lack of opportunity to engage in predatory responses during play can lead to inappropriate predatory behaviour toward human "prey" and owners can find their own ankles and hands being targeted.

Providing appropriate play

Management of these behaviours is achieved by providing appropriate play on a regular and frequent basis and directing such natural responses onto appropriate toys. Fishing rod toys, which stimulate feline interest through the features of rapid unpredictable movement, reflective surfaces and high-pitched or rustling noise are ideal for this purpose. Other items such as laser pointers can also be used to mimic the activity of potential prey but it is important to remember that part of the predatory sequence is the act of catching and killing. Therefore the use of laser pointers, which can never be caught, as the only source of predatory activity is not appropriate and can even lead to the induction of compulsive behaviours which can be difficult to treat. If such devices are to be used it is vital that owners ensure that the laser light frequently falls on an object which the cat can then catch and tear. Suitable objects include so called "prawn parcels" which are made from greaseproof paper and contain food items, such as prawns or chicken. The parcel is tied to the end of a piece of string and either played with independently or incorporated into the laser pointer game. The idea, in either situation, is that the final sequence of the game involves the cat catching the parcel, ripping it open and eating the food inside.

Catering for feline toileting preferences

Hygiene is one of the important features of a natural feline territory and one of the most positive aspects of cat behaviour for many owners is their fastidious approach to toileting. The fact that they are already house trained when they arrive in their new home at just a few weeks of age is something that people find very appealing. However, this reputation for cleanliness leads to high expectations and it is important to recognise that provision of appropriate toileting facilities is a pre-requisite for acceptable and appropriate toileting behaviour.

Location, location

When litter facilities are going to be provided so that cats can urinate and defaecate within the boundaries of the home it is important to ensure that the location provides all of the qualities that a cat will search for in a latrine in the great outdoors and attention to the provision of adequate privacy is essential. Cats will naturally eliminate at the periphery of their territory, away from other resources such as feeding stations and resting areas, and they will select locations where they will be undisturbed. Indoor latrines must therefore also offer seclusion and sites which make the cat vulnerable or force proximity to other resources must be avoided. The fact that cats will often separate the functions of urination and defaecation into different latrine sites when outdoors may lead to a need for separate litter facilities within the home and the provision of multiple indoor latrines which offer the cat choice will usually be beneficial.

Features of the feline latrine

In addition to considering the location of litter facilities it is also important to pay attention to the substrate that is offered within the tray and once again important lessons can be learned by observing the sort of material that cats naturally elect to eliminate in. In general terms preferred outdoor locations offer a soft and rakeable substrate, which can be easily manipulated with the paws during the process of digging an elimination pit and raking over the deposits, and therefore artificial litters should also be selected for these properties. Materials that are hard underfoot or are difficult to rake should generally be avoided since they are likely to decrease the cat's desire to use the tray and may even lead to problems of inappropriate elimination in other locations around the home. Adequate depth of litter must also be considered since cats have an innate desire to bury their excreta and shallow litter may lead to disruption of this important part of the toileting sequence. Likewise litters which are strongly scented are likely to interfere with the cat's natural instinct to bury until the scent of the deposit has reached an acceptable level and use of these products can be a factor in initiating problems of inappropriate indoor elimination.

Ensuring security within the territory

Cats are territorial creatures and marking behaviour is an important tool in ensuring that unnecessary encounters with unfamiliar cats are avoided and territory is kept safe from potential invasion. Both urine and faeces can be used as a form of communication in this context but the deposition of urine as a marker is by far the more common approach and the cat adopts a very characteristic stance when engaging in this behaviour. As a result of using this posture the cat ensures that the urine deposit is ideally positioned to be read by passing felines and that its message reaches its intended target. Such behaviour is perfectly normal and acceptable in the context of the great outdoors and potential encounters with other cats in the neighbourhood are sufficient justification for depositing some clear signals of occupancy around the boundary of the territory. However, marking behaviour is naturally restricted to the home and hunting ranges and whilst keeping the marks fresh is obviously important, in order to signal ongoing occupancy, the number of deposits made by cats that are confident in their right to occupy the location are likely to be minimal.

Decreasing the risk of indoor marking behaviour

Marking within the core territory is not considered necessary and other behaviours such as eating, sleeping and playing are used to identify that area as a place of security and relaxation. An understanding of this use of marking signals is helpful when trying to interpret the inappropriate behaviour of indoor urine marking and by paying attention to the activities that identify an area as core territory it is possible to increase feline confidence and security within the home thereby decreasing the perception that marking is necessary in that context. Obviously identification of the reason for the cat's low level of confidence is going to be essential when managing problems of this nature but when considering prevention of urine marking problems the emphasis is on labelling the house as a place of security and thereby removing the need to mark it. Provision of adequate resources which are readily accessible and of privacy and choice in the home environment will therefore be crucial elements of the prevention programme and the use of pheromone signals (Feliway®) which confirm the familiarity of the location will also be beneficial.

Conclusions

The increasing popularity of the cat as a companion animal has led to a significant alteration in the role that it is expected to fulfil within human society and, whilst the majority of cats adapt well to the challenge, there are others that find the constraints of domestic life difficult to adjust to. Minimising the discrepancy between feline needs and human expectations is therefore essential if cats are to be relaxed in their companion animal role and interpreting situations from a feline perspective will be a significant factor in achieving this goal.

10
Behaviour Problems in the Domestic Rabbit
Anne McBride, Emma Magnus, Georgie Hearne

Introduction

The third most popular mammal pet species in the UK is the rabbit; a species that has been kept as a pet for over 200 years, though it has been kept as a domesticated species for some 2,000 years. Attitudes to this animal have changed quite dramatically since the 1980s as demonstrated by advances in veterinary care, and more latterly pet insurance, in response to demand by owners. Since the 1990s there has been an increase in the house rabbit population and more recently an increased awareness of the role of the pet behaviourist when owners have problems with their rabbit's behaviour. Another reflection of the attitude change is the increasing number of rabbits in rescue societies estimated at 33,000 in 2002, and the increasing number of societies devoted to this single species, currently 300 in the UK alone. Whereas in the past rabbits with behaviour problems would have had a limited lifespan, either being relegated to low level of care until they died, through being abandoned or euthanased.

The rabbit is a common but relatively unknown exotic species. Even simple but fundamental differences of herbivore versus carnivore are not always recognised, with owners providing lamb and rice diets to pet rabbits with diarrhoea as they would their pet dog/cat. Whilst meant with the best intentions, sadly such a lack of knowledge of species can lead to welfare issues both of a physical and psychological nature – as in the case above which did not survive this inappropriate treatment.

What is a Rabbit?

Perhaps of all the common pet species, the rabbit is the most biologically unique. It is not a member of the rodents, as are the gerbils, chinchillas and mice. The European rabbit is sole domesticated member of the Lagomorph family, which includes the hares and jackrabbits. Yet even within this family the domesticated species has interesting and significant characteristics of its own, particularly in relation to the rearing of the young.

The rabbit is an herbivore that evolved in the harsh conditions of the Iberian Peninsular where the herbage is of low quality. It is a selective feeder that spends around 70% of its active time eating. It has an efficient digestive system that includes the production of two types of faeces, soft ceacal faecal balls and hard faecal droppings. The former are caught from the anus and redigested in a process known as refection. This proper functioning of the gut relies on a high fibre diet. The teeth grow continuously throughout the rabbit's life. They are designed to slice and grind thin strips of vegetation, such as grass and hay. Unlike those of rodents, they are not designed to gnaw hard foodstuffs. The provision of such feeds can lead to the mis-alignment of the teeth, formation of sharp spikes, abscesses, and consequently death. The need for high fibre content provided in thin strips which take up a substantial portion of the animal's day to eat is why it is strongly recommended that hay and fresh herbage forms the bulk of the diet, and commercial pellets are provided as a supplement.

In natural circumstances rabbits are rarely seen to drink. This has led to a misconception that rabbits do not need to have a constantly available water supply. However, rabbits like all animals will dehydrate if they get hot or are unable to take in sufficient liquid. In nature this is less likely to happen, rabbits remain cool by being able to go below ground or go into the shade. In addition, feeding primarily at dusk and dawn they obtain a lot of moisture through the dew that has settled on the herbage. It is important to ensure that rabbits always have a freely available source of water.

The rabbit is a prey species, indeed it is one of THE prey species contributing some 20% of the diet to almost 30 different predator species in its native habitat. The rabbit's size means that it provides a neatly packaged, substantial size meal. It has enemies that hunt it at ground level such as fox, cat and dog, other predators attack below ground including these three, but also badger, stoats, weasel, polecat and their domesticated cousin, ferrets. For birds of prey, the rabbit provides a welcome addition to the diet, and young nestling buzzards are fed almost exclusively on rabbits. Of course the predator which allies itself with those that can fly, run or dig and thus catch rabbits is man with his falcons, running dogs, terriers and ferrets, as well as pet cats. Having evolved as Mother Nature's fast food option, it is not surprising that much of its behaviour is orientated towards detection and avoidance of predators.

Rabbits live predominately underground in a complex system of tunnels known as a warren. They spend most of their life in the dark below ground or semi-dark when they emerge between dusk and dawn. This means they are less visible to predators and this also has relevance when rabbit communication signals are considered. One aspect of this is that rabbits do not show obvious signals of pain. This may have evolved because animals that show weakness are likely to be the preferred target of a predator.

The rabbit has acute senses of sight, hearing and smell to enable it to detect potential danger. By living in groups the efficiency of these 'radar' is increased, often with those on 'sentry' duty choosing higher places such as molehills to stand on. The first response of an individual to danger is to freeze, the second being to flee. Only if caught will a rabbit defend itself by kicking and biting. A rabbit will also warn others nearby of danger by loudly thumping its feet on the ground, a noise that travels both above and below ground.

The need to be able to flee to a place of safety is so important to rabbits that they alter their feeding – vigilance behaviour depending on how far they are from cover and how many other rabbits there are around them. Lack of provision of raised sentinel places and cover in the form of pipes and boxes in rabbit pens can result in stress related problems including agoraphobia and aggression.

Rabbits live in groups the size of which is dependant on population density and availability of resources including suitable sites of burrowing and soil stability – the more stable the soil the deeper the tunnels that can be constructed and thus house more animals. Where burrowing is easy the normal group comprises a single female and her mate. Larger groups tend to have more females than males. Groups tend to remain stable during the breeding season (January – August in the UK) with a strong hierarchical structure. This breaks down in the remainder of the year, allowing new individuals to integrate into groups. In general females stay with their natal group and juvenile males migrate to new groups, often having spent several months as satellite individuals whilst the breeding season continues. Male hierarchies are linear during the breeding season which controls access to breeding females and within the group the males co-operate in defence of territorial boundaries, through scent deposition and by deterring intruders. Whilst a dominant female will emerge they tend to have a looser hierarchical structure. This may be due to the fact they are more closely related. Having said this, females will become extremely aggressive over resources, in particular nesting sites, during the breeding season and fights to the death can occur. This sudden and dramatic increase in aggression in early spring is often reported by owners whose animals have lived together peaceably for several months, even the best part of a year.

Rabbits display aggression by an escalating series of signals that are often not easily recognised. As with all species this can be motivated by different emotions.

Competitive Aggression
This often arises between rabbits of the same sex, and less commonly in a domestic situation, of opposite sexes.

As we have discussed, male rabbits are particularly territorial during the breeding season, when they form linear hierarchies within their groups. In an encounter with a trespassing individual, the highest status male will often approach the unknown individual, stopping to restlessly nibble at grass, sometimes engage in some marking behaviours or frantically dig at the ground. If the trespasser does not retreat a fight is likely to ensue. The behaviours likely to be seen before outright aggression include chasing, scraping of the ground and stiff-legged runs past each other. Physical violence rarely occurs at this stage; usually these rituals forestall it. When it does occur it consists of powerful kicks with the hind legs with rabbits often grabbing each other's necks as they kick. Competitive aggression between females usually follows a similar pattern, however it can be more serious.

Defensive Aggression

Aggression of course can be motivated by fear of a threat, commonly known as fear related or defensive aggression. This is a common cause of aggression towards people and again may result from the rabbit's early warning signs having been misunderstood.

When the rabbit becomes frightened the behaviour displayed will usually begin with subtle signals, such as an increase in muscle tension as the animal freezes, its ears will be held back flat to the head and sometimes they will growl and may try to flee. If these signals are ignored the rabbit may well shift its weight onto its back legs, which can give it an appearance of being slightly hunched. Lunging forward and biting often swiftly follows this. This may be a bite and release as the rabbit runs away, or it may hold and kick.

Other Social Behaviours

Given that rabbits spend lots of time in the dark and also try to keep their precise location at any one time hidden they don't display much in the way of visual or vocal communication. This causes difficulties for humans who are used to living with species that communicate predominantly through these methods, such as dogs and cats. They do however communicate socially, although these too are subtle. Affiliative signals include tooth grinding, mutual grooming and seeking close contact.

Scent is the main method of communication that rabbits will use, which originates from three different glands on the rabbit's body:

- Sub-mandibular gland. This is located under the chin, and domestic rabbits will often be seen rubbing their chin on objects, people and other rabbits. This behaviour acts to pass a common scent profile to members of the group and objects within the territorial boundaries and so may well act as a territorial marker.

- Anal-gland secretions. These are deposited with the hard droppings. Rabbits tend to use latrine sites for elimination and all members of the group use these. They are often located on higher ground such as mole-hills and tree stumps where they act as both a visual marker of territorial boundaries from where the scent can be wafted further afield. This behaviour has been used to advantage by the pet owner who can toilet train their rabbit to use litter trays or single areas in the garden as latrine sites. It should be noted of course that rabbits also drop faeces as they forage and whilst the bulk might end up in the litter tray, others will be spread around the environment.

- Inguinal gland secretions. These are deposited with urine, in particular during courtship, and sometimes during territorial disputes. Spraying comprises of the animal running past the target at speed, twisting their hind legs and accurately spraying a jet of urine.

Development

Like puppies and kittens, rabbits are born in an undeveloped state, they are naked, blind and deaf and barely able to move. However, this is where the similarities end. Rabbit maternal care is radically different from that of other species that have undeveloped young.

Rabbit does give birth after a pregnancy of some 28 – 30 days. A fur-lined grass nest is located at the base of a shallow burrow, the entrance of which is blocked with earth whenever the mother is not present – which is most of the time. Rabbit mothers only visit their young once every 20 – 24 hours and then for a mere 3 minutes to suckle them. The young are capable of keeping themselves warm and dry, by eliminating spontaneously once they have fed and then burrowing down into the nest material, only emerging again shortly before the doe is due to return.

Young rabbits will begin to emerge from the burrow in their third week and will be weaned at the end of their fourth. It is then that they start to make their way in rabbit society. The period from birth to sexual maturity is a mere three months (this is delayed in some of the larger domesticated breeds). However, in the wild a female born in the early part of the season is likely to have had one or two litters before the season ends.

It may be considered by some that such a brief time with the parent and short period of development would mean that most of rabbit behaviour is comprised of innate fixed action patterns and more complex developmental issues such as socialising and other behavioural organisation are not relevant, or at least not important. This is not the case.

Studies have shown that handling rabbits between 10 and 20 days of life can have a profound impact on their later willingness to approach humans. In addition, exposure to other species at this age, including potential predators, can reduce a rabbit's reactivity when they encounter these individuals later on. In one experiment, rabbit kittens that had received early exposure to cats did not react fearfully when a cat approached them later on. A similar group, handled only by humans, showed an increased likelihood to approach humans at weaning age and non-handled controls avoided both cats and humans suggesting that rabbits have a socialisation process.

Effects of early handling seem long lasting, as handled females in one experiment raised to adulthood were superior to non-handled individuals in breeding performance. Perhaps one explanation is that early handling reduces stress on the doe, enabling her to raise a litter.

Further research to pinpoint this socialisation period and its profile is required. However the evidence to date strongly suggests that rabbits need to be handled early to prevent later problems. Early handling of kittens should be carried out carefully, ensuring that the hands of the handler are first anointed with the scent of the doe.

Owners need to remember that a rabbit which has been well socialised, and thus has the ability to accept the relevant species, will still remain reactive to certain situations. This is because fundamentally they are a prey and flight species and anything can be perceived as a threat *and cause anxiety and fear*.

Being reliant on flight and escaping to a safe place in a natural environment that is continually changing means that rabbits have to have a constantly updated knowledge of the area in which they live. This in turn means that rabbits have a high motivation for exploration and an ability to learn and remember spatial maps. Rabbits are able to learn through habituation, classical and operant conditioning, as well as by observing the behaviour of others. The principles of training can be applied to rabbits, which can be taught a variety of tricks and cues such as coming when called, fetch and even agility. Indeed rabbit agility is gaining popularity as a sport.

Preventative Measures

Prevention starts with knowledge. Prospective owners need to decide if a rabbit is the suitable pet for their lifestyle and to consider which breed is most apt. The acquisition of a rabbit should be taken no more lightly than that of a cat or dog. Rabbits should not be considered a 'short-term', 'convenient' 'low maintenance' pet. When kept in appropriate conditions some rabbits can live for eight years or more, though for Giant breeds this figure may be nearer five.

Rabbits come in a variety of breeds and sizes, from the tiny Dwarf breeds that weigh around a kilogram to the Giants that can reach the size of a corgi and weigh 10 kilograms. Different breeds also have different temperament characteristics with the smaller breeds generally being more reactive than the larger. This in part reflects the functions for which they were originally bred, the larger breeds having been used for meat or fur production or as laboratory animals, selected for a less reactive more phlegmatic character. Likewise the lop breeds also tend to be less reactive than similar sized more rangy breeds such as Netherland, Dutch, Polish or, somewhat mis-leadingly named, Belgian Hare. Unfortunately, many rabbits are purchased on impulse from pet shops, newspaper adverts or at local breed shows. They are often bought as a child's pet – a pet that often does not maintain the child's interest long term and may then be less than adequately cared for – resulting in physical and behavioural problems.

Having decided on a rabbit and breed, gender is the next consideration. Ideally two rabbits – one of each sex – should be purchased. They should be of similar size and breed, the easiest option is a litter brother and sister. In this way the rabbits will have constant companionship in a natural group, with none of the possible problems associated with introducing strange rabbits to each other at a later date. As soon as possible both members of the pair should be neutered, reducing the likelihood of any seasonal aggression and unwanted youngsters.

Housing is obviously a major consideration and to some extent will be dependent on an owner's preference. Some will wish to have their rabbit as an indoor pet, as one might have an indoor cat. Others will prefer to keep their rabbits outside, preferably in an integrated hutch and run complex. Whichever arrangement is chosen there are some basic principles that need to be adhered to in order to ensure the welfare of the rabbit and continuation of a good human-rabbit relationship.

Rabbits must be provided with sufficient space that they can move freely, stretch to full length, hop, jump and even run. They should be provided with places in which they can hide, take shelter from the sun and rain, stand up on and be away from draughts. Rabbits need to be kept in a well-ventilated environment though being exposed to draughts can result in respiratory problems. Likewise, whilst rabbits will enjoy some sunbathing, they can easily become overheated and must have access to cool, shady areas. After all they are wearing a fur coat that has often provided good insulation to human beings!

Rabbits are curious creatures and will play with objects

9.   Fear of loud noises may cause a dog to seek a location that improves its emotional homeostasis and capacity to cope

10.   Animals may resort to the use of aggession as a means of controlling their environment and regaining their emotional homeostasis

11. Urine spraying – a natural method of communication for the cat, but unpleasant for humans if performed indoors

12. Some multi-cat households are made up of sibling pairs such as these, but in many cases there is no such basis to the feline community and cats are expected to live with total feline strangers

13. Rabbits are curious creatures and will play with and investigate objects in their environment

14. Garden and indoor toys can provide security, shade and stimulation

15. The rabbit's natural behaviour makes it possible to use a litter tray

16a and b.
Creative thinking can provide environmental stimulation for dogs in a shelter environment

Photos: Julie Bedford

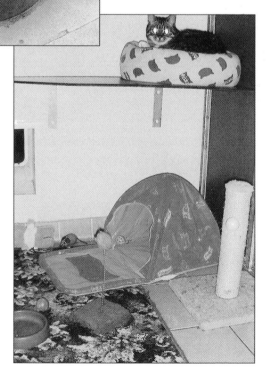

17. Environmental enrichment for the cat in a shelter should include different levels and places to hide

in their environment and should be provided with suitable toys such as the cardboard inner tubes of toilet rolls, toddler teething rings and small footballs. The provision of food should also be adapted to encourage the rabbit to graze. This can be achieved by providing hay in a rack, or within a brown untreated paper bag with some dried or fresh herbs, or threading pieces of green vegetables onto covered wire which can then be suspended from the roof of the hutch or indoor cage. Small branches of untreated fruit trees and leaves can also provide very attractive occupational therapy. There are several toys for rabbits on the market, but not all are totally suitable such as those that encourage the rabbits to use their teeth inappropriately for gnawing.

Last but not least is an appropriate diet comprising mostly hay and fresh greens, with a limited supply of commercial feed and continual access to fresh water. In this way the rabbit will be occupied in a natural manner, namely foraging and grazing, and the likelihood of dental problems, or disease associated with obesity are minimised.

Behaviour Problems

It is only recently that rabbit owners have considered seeking professional treatment for behaviour problems and this means that the area is in its infancy.

The table below outlines four typical rabbit behaviour problems and some of the many possible causes. This highlights the importance of veterinary referral and the need for an extensive case history. Behaviour problems may be multi-causal and treatment rationales will vary from case to case.

Possible Causes	Aggression to other rabbits	Aggression to people	Spraying/ inappropriate elimination (please note there is a difference)	Pica (eating non-nutritional items)
Pain	x	x	x	
Medical issue	x	x	x	x
Poor socialisation and other behavioural organisation	x	x		x x
Inappropriate handling		x		
Hormonal influences	x	x	x	x
Learned response	x	x	x	x
Anxiety response	x	x	x	x
Redirected	x	x		
Territoriality	x	x	x	

To illustrate the complexity of rabbit behaviour problems two case histories and possible differential diagnoses are described in brief below.

Muffin
Pet's name: Muffin
Breed: Mini-lop
Sex: Male – entire
Age obtained: 9 weeks of age
Previous Environment: Outside hutches
Current Environment: House rabbit.
Age: 6 months

Medical History:	Physical examination was unremarkable.
Diet:	Hay, fresh grass and household greens, as well as one quarter of a rich tea biscuit each evening
Month of Consultation:	April

Muffin was obtained at nine weeks of age from a breeder, being a house-rabbit from this age onward. He was litter trained relatively easily, although he will drop pellets around the house occasionally. His daily routine depends on whether there are people home, as he is kept in an indoor hutch rather than left in the house at these times. His owner feeds, grooms and gives Muffin his freedom from 5.30 onwards each day after returning home from work.

The client has sought advice because Muffin urinates on her and 'no-one else'. He will also urinate and defecate on the sofa, and on the client's bed. She has also seen him spray two to three times on the side of the sofa after being interrupted from urinating in his 'usual spot'. In the presence of the owner he will usually visit the site two to three times before eliminating, the owner feels if she is not in the room he will do it immediately (NB: whether this is true is not known, the client leaves the room and on returning there is rabbit urine on the furniture. His behaviour preceding the event is unknown).

The most recent example of Muffin's behaviour was during three hours of free time when the owner's male flatmate was home. The rabbit went into her room for approximately 20 minutes and urinated and defecated on the bed. The previous incident occurred when the owner had Muffin on her lap. She was in a change of clothes as her mother's rabbit had just urinated on her and Muffin climbed on her chest and 'seemed to be searching for more rich tea biscuits'. He then urinated on her. In these situations the owner's response is to shut Muffin away for 15 minutes.

Although Muffin urinated on his owner on the second day she brought him home (at the beginning of December), it stopped around Christmas. Muffin's behaviour started again in February the most recent events were during April.

DIFFERENTIAL DIAGNOSES

It is possible that the lack of social interaction, or company in a gregarious species, particularly when there are unfamiliar scents entering the territory could cause Muffin a degree of stress. As a prey species rabbits are particularly susceptible to the effects of adrenal hormones. As well as physiological illness, behavioural effects of stress include anxiety that can cause over-vigilance and 'over-reaction' to fear eliciting stimuli. Pain, unfamiliar surroundings, loud noise and proximity of predators can all cause stress responses in rabbits. Urination and defecation can both be behaviours that are seen during the 'flight or fight' response and this was considered as a possible cause of Muffin's behaviour.

Physical examination prior to referral would have been expected to rule out many causes of pain including dental disease, spinal fractures or osteoporosis. As his surroundings have remained relatively stable, the firework 'season' has finished and Muffin lives as a house rabbit in a flat with no other pets it is unlikely that his environment is causing him unreasonable anxiety. As Muffin can be held by the owner's flatmate, and seems only to direct his toileting at objects that are associated with the owner herself, anxiety or fear of particular stimuli was considered to be unlikely.

DIAGNOSIS
It is likely that Muffin is displaying normal behaviour in an inappropriate context. It seems to be related to either territorial marking behaviour or possibly an over-attachment or sexual bond to his owner. It is also likely that elements of both could contribute to this problem.

Sooty and Fluffy

Pet's names:	Sooty and Fluffy
Species:	Rabbits
Breed:	Dwarf cross
Sex:	Female – entire, Female neutered respectively
Age obtained:	9 weeks (approximately)
Previous Environment:	Pet Shop
Current Environment:	Hutch & Run with indoor access each evening.
Age:	6 months and 2 years respectively
Medical History:	Routine vaccinations & spay for Fluffy.
Diet:	Commercial mix and fresh vegetables.
Month of Consultation:	March

Fluffy was obtained about two years ago, along with her sister Dotty. Both rabbits lived in the same hutch and cohabited without any incidents. Unfortunately Dotty died unexpectedly at 18 months of age and the family wanted to obtain another companion for Fluffy.

Sooty was bought from a pet shop as a 'six week old male

rabbit' and the initial veterinary check soon after he was purchased confirmed this. From the beginning the family kept
Sooty in a separate hutch and worried that Fluffy was 'intol-
erant' of Sooty. They attempted to introduce the rabbits on
neutral territory; this was the concrete area outside of Fluffy's
hutch. Each time Fluffy would chase Sooty around the pen,
apart from this the children felt that when together the rabbits interacted through a mutual tolerance of each other's
company. In the preceding months Fluffy's chasing has
become more frequent and Sooty has begun to defend herself aggressively. They will now fight on sight of each other.
The children are very anxious about the behaviour but keen
for the rabbits to interact. Observation found they would cry
out and pull the harnessed rabbits apart if there was any
interaction between them at all, anticipating the aggression.

Both daughters sometimes have trouble catching and
handling the rabbits and were keen to improve their relationship with their pets.

Accurate observation and sexing established that Sooty
was in fact female and if put in the concrete area adjacent
to the hutches the rabbits would soon begin to interact,
Fluffy usually chasing Sooty. Both rabbits were on leads
and harnesses and were separated at this point. When in a
grassy run they were less interested in each other and
grazed for around five minutes before passing each other
in their feeding pathway. They were able to be distracted
from each other with dandelion & strips of carrot.

DIFFERENTIAL DIAGNOSES
As in other species, pain can result in aggressive behaviour.
Dental disease or the formation of sharp hooks on the
molars can be extremely painful and rabbits are also
susceptible to developing painful musculoskeletal disorders

such as arthritis or vertebral spondylitis. Whilst both rabbits had been referred as having no medical problems, further medical work up would be appropriate if the situation continued to deteriorate, or if the rabbits start to show physical signs of illness.

A learned pain association may have developed if either rabbit was experiencing musculoskeletal pain and was pulled away from the other rabbit whilst wearing the harness. The children's anxiety led them to pull the rabbits quickly away at the same time as they were investigating one another, which may have caused *the rabbits anxiety through a* conditioned fear response leading to aggression in future. If pain of this type was causing or contributing to the behaviour, it could be that the rabbits were being aggressive when handled or when anticipating handling in the hutch. At the time of consultation they were both able to be picked up, groomed, handled and harnessed without any defensive aggression or avoidance.

It is also possible that limited previous exposure to open spaces or nearby predators has caused fear of novel stimuli in one or both of the rabbits. This type of fear reaction undoubtedly causes a degree of stress that may lead to overvigilance or a decreased threshold for aggression. As both rabbits were kept in a covered area adjacent to the house it was unlikely that nearby cats or birds of prey in overhead trees were causing their behaviour. Similarly both rabbits would graze outside in a run separately and Fluffy spent much time in the area when she was housed with her sibling.

DIAGNOSIS
Most aggressive encounters in rabbits are intra-sexual with females tending to be the more aggressive sex. Sooty's gender was established wrongly at the end of the breeding

season. This is a common mistake as at this time the testes regress into the abdominal cavity. Sooty was purchased in October, her aggressive responses to Fluffy were noticeable in around the beginning of February and have increased in frequency until the time of consultation, suggesting the aggression is due to intra-sexual competition related to the breeding season.

Conclusion

As with all companion animals, the treatment of behaviour problems is not available in a recipe-style format. This is particularly relevant in the field of rabbit behaviour, which is relatively new and the development of our understanding of this species and appropriate methods for behaviour modification is progressing rapidly. All behaviour practitioners, whatever the species they are treating, need a sound knowledge of its ethology, learning theory and its husbandry requirements to complement practical experience. This is also true for the rabbit but people often consider that these are 'easy to treat' animals, or can be treated using similar methods which have gained acceptance in the field of dog and cat behaviour modification. This is not the case and can lead to exacerbation of existing problems or the development of new ones.

11

Rage Syndrome: In our dogs' minds?

Anthony L Podberscek

Aggression in dogs is one of the most common problems treated by behaviour counsellors, and can be categorized in a number of ways, for example, dominance-related, predatory, territorial, possessive and interdog aggression. One type of aggression which is difficult to categorize is rage syndrome. The difficulty arises from our having little knowledge of what actually causes it. To this end, it is usually placed in a category of aggression known as 'idiopathic aggression', which basically means aggression for which there is no known or understood cause. In this chapter we'll look at the current state of knowledge on this perplexing form of aggression.

What Is It?

Rage syndrome refers to sudden, unpredictable, unprovoked and uncontrollable violent aggression towards a person, other animal or object. The term became popular during the early 1980s, when there was a series of newspaper articles (e.g. 'Top Dogs Turn Nasty', *The Sunday Times*, 1981) and magazine articles (eg. 'Solid Cockers Condemned', *Dog World*, 1981) on aggression in English

Cocker Spaniels, in particular those of solid coat colour (e.g. red/golden, black). These were in response to a scientific paper written by animal behaviourist Roger Mugford on the types of aggression in English Cocker Spaniels he treated at his clinic. Indeed, the publicity was such that rage syndrome became synonymous with this breed, leading some to call it 'Cocker rage syndrome'. And this association has since been further enhanced in the media, including a feature article (along with cover photo) in *Dogs Today* in October, 1992: 'Cocker Rage Syndrome: Madness or Make-Believe?' People have even suggested that there are references to a Cocker 'raging' in 19th-century England: that Elizabeth Barrett-Browning, in her letters to Robert Browning, makes references to rage-like behaviour in her male, golden Cocker Spaniel, 'Flush'. So I asked scholar Maureen Adams, who has studied extensively the works and letters of Elizabeth Barrett-Browning, if she knew of any such references to Flush's behaviour. She provided me with the following excerpts from Elizabeth Barrett-Browning's letters:

'Critics who bark the loudest commonly bark at their own shadow in the glass, as my Flush used to do long & loud before he gained experience & learnt the [in Greek] 'Know Thyself' in the apparition of the brown dog with glittering dilating eyes#"1993, August 8, 1845

'For Flush, though he began by shivering with rage & barking & howling & gnashing his teeth at the brown dog in the glass, has learnt by experience what that image means, . . . & now contemplates it, serene in natural philosophy' 2271, March 24, 1846

Also, Maureen found that Flush did bite Robert Browning on two occasions.

While these are interesting accounts of Flush's behaviour, it's unlikely that they are indicative of rage syndrome. Indeed, Maureen found that Flush was very protective of Elizabeth Barrett-Browning and believes that this may have been why he bit Robert Browning.

This is not to say, though, that Cockers do not suffer from rage syndrome, rather that the breed seems to have received undue attention in relation to it, and that the implication is that other breeds do not suffer from it. Fortunately, not all publicity for the English Cocker Spaniel has been bad, though. Scamper, the golden Cocker in Enid Blyton's *Secret Seven* books, had an impeccable temperament, apparently!

What Happens?

From the outset, it is very important to stress that the information provided here is based on the limited amount of information available on the topic, and that in many cases a diagnosis of 'rage syndrome' has been based purely on behavioural signs alone.

So, what happens? Owners report that their dogs are normally healthy and good-natured, but that from time-to-time, and without any warning, the animal will become violently aggressive (biting – barking may occur as well), usually towards the owner or other member of the household, although attacks towards another dog or an object have been recorded in the some cases. During the attack the dog's eyes appear glazed and its pupils are dilated, and some people have reported that the eyes appear red in colour. In addition, some dogs drool, vomit, defecate and/or urinate during an attack. By all accounts, the dog does not seem to know what it is doing and its behaviour is uncontrollable. Attacks towards people seem to be provoked by very simple and seemingly inconsequential interactions, for example, the owner bending down to pat the dog, the owner brushing past the animal, or the owner giving a command in a normal and friendly manner. Sometimes, just before an attack, owners have noted that their dog's behaviour changes; the dog appears to be in a trance – it stares or growls at nothing in particular. In other cases, attacks occur soon after the dog has just woken up. After an attack, which usually lasts for up to five minutes, the dog appears to be unaware of what has just happened, is tired/exhausted, and usually goes off to sleep for 30 minutes or more. Afterwards, the dog appears normal and friendly again.

This problem has been recorded in dogs as early as six months of age, but most reported cases involve dogs from one to three years of age. Male dogs are most commonly implicated, but it is not possible to say anything about the neuter status of these dogs, as it has not always been reported in the literature.

Interestingly, the dog in question often occupies a high status position in the owner's home (a lack of inhibition is present in the relationship with the owner leading to agonistic behaviour) and it may also display territorial aggression.

Breeds of Dog Involved

As mentioned earlier, rage syndrome has become synonymous with the English Cocker Spaniel. However, a number of other breeds have demonstrated this condition, or at least signs indicative of it. These include: American Cocker Spaniel, Bernese Mountain Dog, Bull

Terrier, Chesapeake Bay Retriever, Dobermann, English Springer Spaniel, German Shepherd, Golden Retriever, Jack Russell Terrier, Lhaso Apso, Pyrenean Mountain Dog, Rottweiler, St. Bernard, West Highland White Terrier and Yorkshire Terrier. It's important to note here, though, that in most cases only one dog of each breed has actually been reported on in the scientific literature, and usually only behavioural signs have been used as evidence.

What are the Potential Causes?
There are a few theories as to what might be causing these outbursts of sudden aggression in dogs.

STATUS-RELATED AGGRESSION
Although owners will claim that they did nothing to provoke the attack on them, behaviour counsellors have noted that these people often have a problem with status-related aggression with their dog; that the dog sees itself as having a higher status in the household than the owner. In these cases, aggression can occur when the owner inadvertently does something that the dog doesn't like. These attacks will appear unprovoked because they are often simple, seemingly innocuous things, such as the owner going to pat the dog, to step over the dog, or to take something which is near the dog. Indeed, my research with James Serpell on 1,109 English Cocker Spaniels showed that those which showed aggression 'suddenly and without apparent reason' also often showed signs of status-related aggression (eg., aggression when disciplined, aggression when reached for or handled).

It is also important to note that different breeds of dog, and indeed different individual dogs, will have different thresholds for aggression. So it's possible that a breed such as the Cocker Spaniel has a low threshold for aggression, and this may be partly why it has attracted more negative media attention than some other breeds of dog.

COMPLEX FOCAL SEIZURES; BEHAVIOURAL EPILEPSY
This is a type of epilepsy which manifests itself in the form of behavioural change, rather than the classic fits/convulsions we usually associate with epilepsy. Epilepsies can be broken down into two main groups: primary (idiopathic) and secondary. The primary epilepsies are those for which there is no identifiable brain abnormality other than the seizures themselves; the cause is unknown. Secondary epilepsies are those in which seizures are the consequence of an identifiable lesion (e.g. tumour) or other specific cause. Primary epilepsies can be further broken down into generalized seizures (involving both hemispheres of the brain) and focal seizures (abnormal activity in one region of a hemisphere of the brain). Generalized seizures include the ones most of us associate with epilepsy: loss of consciousness, convulsions and foaming at the mouth. Focal seizures can be either simple (consciousness is preserved during the seizure) or complex (consciousness is *impaired* during the seizure). We are interested in the complex ones, as this type of epilepsy can manifest itself as behavioural change (e.g. aggression) with impaired consciousness, general signs in common with rage syndrome. Other behavioural changes which have been observed in dogs during a complex focal seizure include sudden fearfulness, barking, 'fly-catching' (snapping at the air as if trying to catch flies) and tail chasing. In some cases, vomiting and diarrhoea have also been reported during the seizure. In

cats, complex focal seizures manifest themselves in the form of confusion, defensiveness, aggression, hissing, moaning, and involuntary licking and chewing movements.

Focal seizures are thought to originate from a variety of places in the brain: the temporal lobes, limbic system, frontal lobes or the occipital lobes. If the location of the origin of the epileptic seizures is known, then a more specific diagnosis of the condition can me made; eg., temporal lobe epilepsy.

When aggressive behaviour is a feature of an epileptic seizure, typical characteristics of seizures must accompany the aggressive act. That is, the seizure should begin suddenly and without provocation, last for a brief period (approximately one to three minutes) and end abruptly, with the violent act occurring in a context of impaired consciousness (dog is disorientated/ unresponsive).

And while most seizures appear to occur spontaneously, they may be precipitated by a variety of factors. For example, sleep deprivation, emotional stress and even heat in unspayed female dogs can precipitate a seizure. The most common type of trigger in people is flickering light, usually from a TV. Certain sounds can also act as a trigger.

However, although aggression during an epileptic seizure is possible, it has not been frequently documented in the medical or veterinary literature. And yet associations between epilepsy and violence prevail in the popular press and among scientists. A major problem is that some scientists and practitioners fail to distinguish between aggressive behaviours that are directly linked to seizure activity and aggressive behaviours that are related to other factors, such as the presence of other underlying brain abnormalities.

NONEPILEPTIC SEIZURES

These are seizures which appear to be epileptic in nature (e.g. may appear to be a complex focal seizure), but in fact are not due to epilepsy. Instead, other types of brain dysfunction/disturbance are involved. In the medical literature, people demonstrating uncontrollable rage attacks, which are not due to complex focal seizures, are thought to suffer from intermittent explosive disorder (sometimes also referred to as episodic dyscontrol syndrome or rage). This is characterized by repeated acts of violent, aggressive behaviour in otherwise normal people, and the behaviour is out of proportion to the event that triggers or provokes it. Attacks are followed by fatigue, amnesia and sincere remorse. Sufferers are usually male and it is thought that problems in the temporal lobe/limbic system areas of the brain are the cause.

The problem we face initially with diagnosing what is actually causing the sudden aggressive outbursts in dogs is that the behavioural signs are common to all three causes detailed above. Without further investigation, we cannot make a definite diagnosis.

Diagnosing the Condition

The challenge is being able to diagnose the problem correctly; a full medical, behavioural and neurological examination is required. It is important to remember, too, that aggressive outbursts may be the result of something other than rage syndrome, for example, brain tumours, toxicity (e.g. lead poisoning), infectious/inflammatory diseases, head trauma and metabolic disorders (eg., low blood sugar).

I believe that rage syndrome should only be used to refer to *nonepileptic* behavioural seizures which manifest themselves in the form of sudden aggression. In people, these types of seizures have been labelled intermittent explosive disorder, episodic dyscontrol syndrome or rage attacks. When we have evidence that the aggressive outbursts are actually epileptic seizures, then the diagnosis should be behavioural epilepsy/complex focal seizure. But just to complicate things a little, human sufferers of nonepileptic behavioural seizures sometimes also have epilepsy! So we need to consider the possibility that a dog showing behavioural seizures may be suffering from both rage syndrome and epilepsy. Where there is no evidence of behavioural seizures (epileptic or nonepileptic) a diagnosis of status-related aggression would be more appropriate – this is not rage syndrome.

So how do we find clinical evidence of behavioural seizures? To date, the best diagnostic tool is an electroencephalogram (EEG). This is a recording of the electrical activity of the brain, abnormalities of which can indicate a variety of brain disorders, including epilepsy. Complex focal seizures always show an abnormal EEG *during* a seizure. For nonepileptic seizures, the EEG will always be normal during a seizure, while non-specific abnormal activity occurs *between* seizures.

How Common is the Problem?

To date, there have only been a handful of suspected cases which have been fully documented in the scientific literature. In addition, reports from both UK and USA behaviour counsellors suggest that very few cases have been seen in clinics. From my research on English Cocker Spaniels, it was found that approximately 8% of these dogs regularly showed aggression 'suddenly and without apparent reason'. An earlier Dutch study on 404 Bernese

Mountain Dogs found that approximately 20% of them showed 'intermittent attacks of blind aggressiveness towards their owners'. However, these were both questionnaire-based studies, and full behavioural and medical/neurological assessments on these dogs were not conducted. Consequently, we can't say anything about the dogs in terms of how many actually suffered from rage syndrome. Indeed, from my research on English Cockers, most of the potential ragers also showed behaviours indicative of status-related aggression.

What we do know is that seizures are the most common neurological problem reported in dogs: approximately 0.5 to 5.7% of dogs have experienced seizures sometime in their lives. This is most likely an underestimate, though, as many people do not know that seizures may manifest themselves in the form of behavioural change and because owners may not bother to take their dog to their veterinarian if the seizures are minor or infrequent. A recent study on Poodles and Dalmatians who had seizures showed that the majority suffered complex focal seizures, although none manifested themselves in the form of sudden aggression. The behavioural manifestations observed were 'fly catching' and sudden fear.

Can it be Inherited?

There certainly is potential for this because, based on pedigree analyses, a genetic basis for primary (idiopathic) epilepsy in several breeds has been shown. The breeds included are Beagle, Belgian Tervuren, Keeshond, Dachshund, German Shepherd Dog, Labrador Retriever, Golden Retriever and Collie. As to whether complex focal seizures manifesting as aggressive behaviour are heritable, the jury is out, as not enough studies have been conducted.

From my own research on English Cocker Spaniels, it was clear that solid-coloured Cockers (red/golden, black) displayed most types of aggression, including 'suddenly and without apparent reason', much more commonly than did parti-coloured Cockers (eg., blue roan, black & white). And because solid and parti-coloured Cockers come from distinct bloodlines, this is a pointer to there being a genetic basis for the difference. However, this is not evidence for rage syndrome being genetic, as it was not possible to determine which, if any, of the dogs in the study actually suffered from it.

How to Treat the Condition

If a dog is found to be showing only status-related aggression – and therefore is not rage syndrome – a behaviour counsellor can advise the owner on how to change their behaviour to bring about a more balanced and harmonious dog–owner relationship, with the owner firmly in control. The chances then of aggressive outbursts from the dog should be very greatly reduced or even extinguished.

For dogs suffering from epileptic and nonepileptic behavioural seizures, manifesting themselves as sudden aggression, anticonvulsant medications such as phenobarbitone, primidone and potassium bromide are the drugs of choice. Nonepileptic seizures may also be controlled using other mood-stabilising drugs, such as clomipramine. In addition, if specific triggers (certain sounds, owner behaviour) for these seizures can be identified, then it may be possible to advise the owner on how to avoid them.

Conclusion

At the end of the day, we know very little about rage syndrome. And yet, unfortunately, many people use the term too readily when diagnosing dogs which show unpredictable aggression, especially if that dog happens to be an English Cocker Spaniel. Without conducting full medical, behavioural and neurological assessments, no practitioner can be very certain as to what they are dealing with or how best to treat it.

In the veterinary field, there is a long way to go to understanding and diagnosing neurological problems such as rage syndrome. It's a fascinating area which deserves *much* more attention.

Acknowledgements

I am extremely grateful to Maureen Adams, Laura Marsh, Gail Podberscek and Adrienne Thomas for their help.

12

Rehabilitation of Rescue Dogs and Cats

Julie M Bedford

Two common misconceptions about animals available for rehoming at the many charity animal centres is that they have been rescued from a life of neglect and abuse or have behaviour problems that make them too dangerous or disruptive for the average person to manage. Thankfully neither is usually true. Most animals admitted to charities for rehoming are well cared for and much-loved pets, which when obtained were intended to be with the family for life. In fact the most common reason for animals being handed over to one rescue organisation, The Blue Cross, is changes in the owners' circumstances. These include changes in work pattern, financial difficulties, divorce, house moves and illness/hospitalisation. The second most common reason for animals being brought to The Blue Cross is that they are strays. Almost 10% of dogs and 30% of cats handed into The Blue Cross are strays.

The least common reason for animals being handed to the Blue Cross is behaviour problems. This group accounts for less than 10% of cats and 9% of dogs. Of those, most will have their behaviour problems resolved or managed. The majority are overcome by providing a more suitable activity, a change of environment or by placing the animal with an owner who does not see the behaviour as a problem due to their experience or expectations. So the majority of pets available at rehoming centres will be problem free and, with correct placing and introduction to the new home, will settle successfully.

What follows provides an insight into what is involved in the rehabilitation process, which is a guide for those working or aspiring to work within a rescue organisation. For the pet owner or potential pet owner it is an insight into the care taken where best practice is employed.

Of the remaining few most of them can be helped with sensible handling and a sensitive well informed approach to their management and rehabilitation.

The rehabilitation process, for those that need it, has to begin when the current owner first approaches the animal charity to hand over the pet. Prior to, and at the hand over as much information as possible should be gathered. Comprehensive information helps to define the problem and the appropriate way to resolve it.

Taking a Full History

When the animal is brought to the centre for rehoming it is important that a detailed account of the animals routine, management and behaviour are taken. This must be done without judgement – generally people do what they think will be the best to resolve their problem and taking a judgemental stance will prevent the client being honest and open. The client must be reassured and forewarned of the need to complete a questionnaire about their pet. The person collecting information should be genuinely interested in getting as much detail as possible but must do it sensitively and professionally. This process must be followed whether or not the animal has behaviour problems. The more information that is taken on an animal the more likely it is to be successfully rehomed.

Following is a list of questions that should be asked – the list is taken from the admittance form used by Blue Cross Adoption Centres. The questions are only a starting point and should be developed and explored as the initial answers are given. Questions are better kept open (avoid yes and no answers) to ensure descriptive and qualitative information.

GENERAL QUESTIONS THAT ARE ASKED OF DOGS AND CATS

AGE

SEX (NEUTERED)

BREED

TEMPERAMENT DETAILS

THE FOLLOWING GIVE A GOOD INDICATION OF THE OVERALL BEHAVIOUR OF THE ANIMAL:

How does he behave with strangers (men and women)?
How does he behave with children (0-4years, 5-12 years, teenagers)?
How does he behave with dogs, cats?
How active is he?
How strong-willed is he?
How affectionate/tactile?
How does he behave when restrained/groomed?

Is he frightened of anything (noise, vets, cars etc)?
How does he behave at the vets?

GENERAL MANAGEMENT DETAILS
How long can he be left without problems?
Does he travel well?
Is he housetrained?
When and what is he fed (try and keep the dog on the same food and gradually change over to the brand used at the centre)

HAS THE ANIMAL BEEN SOCIALISED
How old was he when you obtained him?
Where did he come from (breeder – kept outside, domestic setting, etc)?
Who lives in the home (adults/children/pets)?
How many homes has he had (including yours) – if he has had previous homes why did the previous owners give him up?
Is your home quiet, busy?
As a puppy/kitten did he meet many adults, children, dogs, cats did he have any unpleasant experiences with any of them?

THE OWNER/ANIMAL RELATIONSHIP AND WHAT TYPE OF INTERACTION THERE HAS BEEN
Have you ever needed to punish your pet – if so how did you do it and what were the circumstances?
What makes your dog/cat happiest – list three things
How responsive is he to commands?

WHY ARE YOU GIVING YOUR PET UP FOR REHOMING?

DOG SPECIFIC QUESTIONS
Did the dog have a lot of opportunity to play with other dogs when growing up?
How did he play with the other dogs? Eg., which dog initiated play, describe how they played etc.
Where does the dog sleep at night?
Is he allowed on furniture?
Can you take bones, toys, and food away easily?
What is your dog's favourite game – tug of war, chasing, squeaky toys, won't play with toys, wrestling games with people?
How many times is your dog walked? Is he off lead? How long are the walks?
Has your dog ever growled, barked, snapped or bitten strangers (including visitors to the house) or members of the family?
What were the circumstances?
Does your dog follow you from room to room?
How does your dog try to get your attention?
What does your dog do when left alone?

CAT SPECIFIC QUESTIONS
Is the cat . . .
Confident?
Sociable?
Shy?
Does he prefer to be alone?
Can you give medication to the cat easily?
Does the cat like to be picked up?
Does he enjoy being cuddled, stroked?
How does your cat behave with cats outside the home?
How many other cats are in the home – how does he interact with them?

Does he use a litter tray (if so what litter is used etc)?
Is this shared with other cats?
Does he have a scratching post – does he use it?
Is he used to going outside?
Can he use a cat flap?
Has the cat been used to living near a busy road?

The answers to these and follow on questions are used to build up an individual profile. In addition to providing a guide as to the most suitable type of new home for the animal it can be used to identify problem behaviour, how and why it developed and how the owner dealt with it. A plan can then be developed as to how to better manage the animal and a programme devised to avoid or resolve the problem.

The information given by the owner is very important as it indicates how the animal behaves in a domestic environment, what it has learned and how it has been managed. But it must be supplemented by observations by staff whilst the animal is at the centre. Ongoing assessment and keeping a record of the initial and ongoing information where all staff dealing with the animal have access to it is essential for successful rehabilitation and rehoming.

Ongoing assessment
This is important for all animals but particularly so for those with no known history eg. strays.

The ability to make predictions from assessment in environments such as kennels and catteries is limited and particularly difficult in cats (whose behaviour tends to be suppressed in new surroundings). Ongoing assessment can help to support the profile and identify problems that may require some work. To be valid the assessments cannot be done as an intensive set of kennel or cattery tests and observations (of the type where people introduce novel items or mildly intimidate to see what response they provoke). Although helpful, formal testing of this kind is artificial, in respect of context and because they are a snap-shot of the animal at a particular moment, tiring and stressful and does not yield reliable results. Assessments are better and more realistic if carried out by appropriately trained staff during the normal routine of handling and caring for the dog or cat. For example noting how the dog behaves when out for a walk or how well the cat tolerates grooming builds up a more accurate picture of the animal's responses and ability to cope with new situations. These 'real life' situations and are far more useful and predictive than contrived 'formal' tests.

Minimising the Effect of the Kennel or Cattery Environment

The contrasts between a home environment and a kennel or cattery is immense and seeing how well most animals cope with the transition is a testimony to just how adaptable and robust most animals are. However some animals' behaviour will deteriorate whilst in kennels and catteries. This can be minimised by careful attention to their routine, environment and handling. A routine gives the animal pre-dictability of 'what will happen when' and add to the feeling of security. The kennel or cattery design should enhance this security and provide an area for the animal to retreat to if they are feeling overwhelmed or threatened. Physical comfort should also be ensured, temperature should be appropriate for the animal, the housing should be dry and draft free and noise kept to a minimum. Ensuring the dogs have a routine so that barking is restricted to times of excitement, eg. feeding times, can reduce noise in kennels. Radios should not be too loud or left on all day.

Environmental enrichment should provide for the animals' behavioural needs. For example, cats need scratching posts, boxes to hide in, toys and high shelving. Dogs need toys; chew items and many enjoy extras such as paddling pools. Dogs also benefit from regular walks in a variety of different areas – pavements near traffic, fields near livestock. These activities not only contribute to the assessment of each dog but also prevent them losing their familiarity with a variety of environments. Handling should be done daily, should be non-intrusive and gentle. Gradually these sessions should be built up to full grooming sessions and 'vet' like inspections. With basic routines and handling methods in place and consideration given to the animals' environment the next step is to work out what rehabilitation work needs to be done.

Making a Plan and Ensuring Everyone is Working Towards It

When an animal is identified as needing rehabilitation it is important to discuss and agree a plan of action for that dog or cat. Staff meetings when everyone can feed back their experience with the animal are a useful starting point. A plan of action must be agreed and then adhered to by all staff. Reminders such as notes on the animal's accommodation must be clearly posted. Progress reports must be kept to ensure each individual takes responsibility in carrying out the programme. Progress should be regularly reviewed to ensure the rehabilitation programme is improving the animal's behaviour.

As members of the public usually have access to the animal areas they may need to be advised as to how to behave around a particular animal. If constant supervision is not possible then instructions must be clearly displayed. For instance an animal that is intimidated by people staring at it may have a sign displayed asking people not to stare – and most importantly why they should not stare. An alternative behaviour must also be given such as looking past the animal and speaking softly. If people understand why they have been asked to do something, and given an alternative behaviour, they are much more likely to comply and not hinder the progress of the dog or cat.

The public can even form part of the rehabilitation plan. Where animals are timid but not aggressive a small holder can be attached to the front of the kennel or cattery and filled with food treats. A notice can be attached asking people to throw a tit bit in as they pass. This helps the animal associate strangers with a treat rather than intrusion. With young or boisterous dogs this can be taken further. The food treat should only be given if he is sitting quietly by the door (this should be reinforced by staff before relying entirely on the public). This is surprisingly effective and engages the public as well as the animal.

Having fishing rod type toys available in the cattery encourages the public to spend time playing with the animals in a non-intrusive way and can be done through the mesh of most cattery fronts. In addition to encouraging the animal to form pleasant associations with strangers these methods also assist in keeping the animals occupied.

Occupying the Animals, and Training in the Centre

Bored and under stimulated animals are likely to invent ways of occupying themselves. Usually the things they find to do are not what pet owners want. Cats and dogs both benefit from being taught how to play with toys. Not only does it channel their energy and provide an outlet for some of their natural behaviours, it also assists in building bonds between the animal and people. Keeping the animals occupied and breaking up their day gives them something to look forward to and helps prevent their behaviour deteriorating. Feeding dogs at least twice a day helps them utilise their food more effectively but again adds an extra positive encounter with their handler and an opportunity to train eg. asking the dog to sit before feeding. Feeding dry food via a food ball is something that cats seem to particularly benefit from. They are kept active 'chasing' their food and this outlet for their normal hunting behaviour also appears to reduce some cats' aggression towards their handlers.

Short, regular, handling sessions for both cats and dogs is particularly useful – these should be pleasant and done at times other than those used to give medication etc.

Basic training can also make a big difference – particularly for dogs. Teaching good manners, self-control and how to work for rewards not only makes a dog easier to re-home it channels their energy, develops alternative behaviours that can be rewarded and prevent reinforcing unwanted behaviour. Basic exercises include waiting by doors, not jumping up, walking on a lead, sit, down and coming when called. Training must be consistent and implemented from the time the dog first arrives at the centre. Rewarding the dog whenever it displays the required behaviour will ensure it will quickly learn what behaviours you want and which behaviours don't bring it praise, food or games. This need not take any additional time as it can be incorporated in the dog's daily walk. Walking dogs that are kennelled is essential – not only does the dog get attention on a one-to-one basis and a break from the kennel but as mentioned earlier it is an ideal opportunity for the handler to assess the dog and note how it behaves when it comes across traffic, strangers and other animals.

Minor Behaviour Problems

It is always important that any medical conditions are ruled out but problems not involving aggression can often be dealt with very effectively in the kennel or cattery. For example, attention-seeking problems are often resolved in rehoming centres as people are just not available to respond to behaviours exhibited to get attention. Providing this type of problem has been identified and the new owners advised when to give attention and when to withhold it the problem is quickly resolved.

Spraying by cats, often caused by stressors – such as too many cats in the home, dogs, lots of visitors etc can be resolved by correctly identifying the problem (from the questions taken when admitted) and rehoming to an environment in which the cat is more likely to cope. For example, a quiet home with no other pets.

House training problems can be resolved by providing the cat with a suitable covered, private litter tray and ensuring that a similar litter is used in the new home, placed in a quiet area easily accessible to the cat. Dogs that have not been house trained or their house training has broken down can be re-homed with advice on appropriate house training methods and ensuring that the dog has been well exercised and toileted before going into the home for the first time. Similar advice will also prevent the breakdown of house training, as many animals are excited or nervous at being taken to a new home – in the excitement they may urinate or defecate in the home, especially if they have had a long car journey. This marks out for them a new 'toilet area' and in a new home they only way they can identify where to toilet is by smell.

It is also important to stress to the new owner to start off as they intend to continue. Many owners feel sorry for their new pets and indulge them in behaviours that may not be acceptable later on in the relationship. For example, dogs may initially be allowed on the furniture, but when they are settled in or when the weather is wet the owner decides that they can no longer be on furniture. Having to learn new rules is confusing for the dog and may cause conflict. It is better for all, speeds up the settling in process and prevents problems developing, if the animal has consistency from day one.

Nervousness

Nervousness is a common problem in dogs and cats that are in rehoming centres and it often stems from inadequate early life experience. Whilst that can never be fully overcome a gentle desensitisation programme will start to help the animal readjust and cope with life. It is important that nervous animals are kept safe and relatively quiet – so place cats well away from the kennels and keep nervous dogs in quiet areas of the kennels. Avoid over-stressing the animal by having multiple handlers or intrusive practices. Initially just one or two members of staff should care for the animal. Allow the animal to build a bond with them before introducing new people. Basic considerations are:

- Handling the animal at times other than when needing to do intrusive procedures such as worming and vaccination.
- Giving the animal a sense of control in its ability to move away and feel safe. So cats need a hiding place and

high up shelving. Dogs need an area they will not be stared at (simply achieved by throwing a blanket over the front of the kennel door so it is partly obscured and the dog can hide behind it). Animals that know they have a safe retreat are much more likely to come forward and investigate the visitor.

- Trying to link the people and things the animal is afraid of with something the animal enjoys – so a stranger may engage the animal in play or offer a particularly tasty food treat, as discussed above.

- Allowing the animal to approach rather than approaching the animal – this takes longer initially but by going at the animal's pace it will help the animal to develop his confidence and learn to trust people. Progress is then much quicker.

- Teaching the animal alternative behaviours, eg. giving a paw, lying down, etc, is also useful – nervous and fearful animals really appreciate being given a different way of coping – but only if you ensure they are not threatened or overwhelmed. If the animal does not trust you it will have to rely on old behaviour patterns to cope with new situations.

- Training a new behaviour is something that is often considered and done with dogs but rarely with cats. Cats in particular respond very well to clicker training and now that clicker training is becoming more popular and more people are becoming competent at using it, it is an ideal method for re-training many cats.

Aggression

Rehabilitating aggressive animals depends on the type of aggression, how long the animal has been aggressive and how severe the aggression is. If staff, public or other animals are seriously at risk then the decision may be taken not to attempt rehabilitation. However many types of aggression may be avoided and controlled by careful rehoming and advice.

Some aggression, particularly in older cats, may be attributed to pain. After veterinary examination careful handling and treatment for painful conditions such as arthritis often transforms the cat's behaviour.

Cats aggressive towards other cats in the home can be rehomed singularly and in an area where there are few other cats. (Despotic cats are an exception as they may travel considerable distances to find and attack other cats). By not being forced into close contact with other cats the cat will be able to avoid encounters with them and general aggression levels will fall. Cats may also be aggressive towards their owners – this can be for a variety of reasons but most can be reduced by ensuring the cat has sufficient stimulation – access to outdoors or appropriate play with a fishing rod type toy (these are particularly good as the owner can keep a safe distance from the cat). Owners can also be taught to recognise early warning signs and management techniques. Cats that bite or scratch when picked up can be taught to jump on the owners lap by luring up with a prawn or tuna fish. Watching out for body language and facial expressions that accompany aggression, such as twitching skin, tail flicks and body stiffening will enable the owner to move before the cat tries to move them.

Dogs that are aggressive against other dogs can become much worse in kennels but if there is the option of isolating them (for example keeping him in the staff room) and teaching him to bond strongly with humans then a great deal can be done at the centre. When the dog has developed a strong bond with people it can be gradually exposed to the other dogs in a controlled and safe way. Additional control can be achieved by using a head collar that will give the handler the opportunity to divert the dog's attention and prevent him lunging towards the other dogs.

Dogs that are accustomed to controlling their owners and demanding what they want and have become uninhibited can be sent out with modification programmes advising how to manage the dog and ensure it works for rewards and is not placed in a situation where it can guard resources.

In all cases follow-up behaviour advice and support can help the new owner to resolve the problem (the new owner must be fully informed of the problem prior to adoption). It is the ongoing support that ensures the owner is motivated and confident enough to resolve the problem.

Careful Selection and Matching of New Owners

Matching animals with new owners takes a lot of time and effort, but not doing it is a false economy. Getting the right match saves a great deal of time, as the same animals aren't returned for 'recycling'. It is better to refuse a prospective owner an animal that is not suitable and keep it for a while longer, than place it just because it is better in a home rather than a rescue centre. By fully explaining the decision to the perspective owner and identifying a more suitable animal for them you have a better-educated owner and increased the chance of the adoption being successful. However, this policy must be flexible – most of our pets are not ideal for our circumstances because we accommodate their needs, so again individual attention is important.

Of course it is important that we do everything to rehome correctly the first time around. Failing to do so costs time and money which charities can ill afford and more importantly has a negative impact on animal welfare. Moving from home to home and being brought back into kennels/catteries – however well run is unsettling and distressing for dogs and cats. In all cases follow-up behaviour advice and support can help the new owner to resolve the problem.

Strays
The history of these animals is unknown and additional precautions will be needed when rehoming. For example it is wise, in order to prevent problems developing, not to rehome adult animals to homes with young children. This is because the dog or cat's experience with children is not known and it is unfair to the animal and family to risk placing a dog or cat into what could be a very difficult environment for that individual animal.

Ongoing Support and Advice
The unsettling experience of changing homes and the different expectations of different people can lead to behaviour problems developing which the new owner will find problematic. These cannot always be foreseen but if the client has been made to feel comfortable during the adoption process and information has been freely available they are much more likely to seek and value advice given by the rehoming organisation. The new owners are also much more likely to seek help as soon as the problem becomes apparent rather than waiting for it to become unmanageable. In the early stages most behaviour problems can be explained and resolved by giving good quality

information via the telephone or information leaflets. Some problems, especially those involving aggression will need much more input, including regular visits and an individually prescribed modification programme. Taking this sort of action in good time can prevent new owners having to return the animal.

Taking an interest in each animal and person arriving at a rehoming centre, gaining and passing on as much relevant information as possible about them will ensure the best possible match is made. This combined with ongoing support in the form of good quality information, openness, help and advice ensures a high standard of animal welfare and customer care. Working towards these principles enabled The Blue Cross animal welfare charity to have less than 4% of dogs and 1.5% of cats, rehomed by the charity, to be returned.

13

Behavioural Problems in Ageing Pets

Sarah Heath

Introduction

As the state of health care continues to improve more and more of the human population are living long enough to have to deal with the varied implications of old age and society is recognising that the ageing population has specific needs. In a similar way the pet population is steadily growing older and the proportion of pets that qualify as geriatrics is increasing. With this change in population comes a need to recognise the specific requirements of this special group of animals, in terms of their behavioural and emotional needs, and to offer appropriate support for their owners.

An Inevitable Process

Ageing is one of those inescapable facts of life and veterinary surgeons are no more able than their medical counterparts to offer any miracle solution to the issue of old age. However, there is now a considerable body of information available about the changes that occur during the ageing process and about how these changes can be affected by diet, lifestyle and medication. With this knowledge at our finger tips we can certainly increase the quality of life for the ageing pet population and ensure that pets and their owners get the most from their relationship at this time.

A Changing Population

Advances in veterinary medicine over recent years have meant that old pets are now offered the very best in medical care during the later years of life and many will enjoy excellent physical health into their late teens and even beyond. However it is also important to remember that ageing affects not only the major organs such as the heart, lungs, liver and kidneys but also the nervous system and more specifically the brain. Changes in neurotransmitter levels, alterations in membrane permeability and increased production of free radicals within the central nervous system can all lead to the onset of age related behavioural changes and when pets begin to show symptoms related to these changes they begin to require a very specific form of veterinary attention. Of course behavioural changes in elderly patients are not always related to specific issues of brain ageing and the potential involvement of medical conditions such as pain, failing sensory systems, compromised blood flow to the central nervous system and age related deterioration in functioning of the major organs, such as liver, kidneys and heart, all need to be considered. In addition environmental and psychological factors should not be overlooked and it is important to remember that older animals may be less adaptable to change in their social or physical surroundings and may show marked behavioural changes in response to changes in their household or routine as a result.

Slowing Down or True Brain Ageing?

As the ageing process takes its toll on a dog's heart and brain some changes in behaviour and personality are almost inevitable and it is important to be able to differentiate between cases where the animal is simply slowing down and those where the animal is finding it increasingly hard to function at a social level. In some cases behavioural changes in old dogs will noticeably resemble the symptoms of senility or Alzheimer's disease in people. Lack of connection between behaviour and context is a classic sign of brain ageing and many owners have reported that their pet seems like a stranger in its own home. In cases of cognitive dysfunction in dogs and cats post mortems have shown similar neuropathological lesions to those seen in people with Alzheimer's disease and there is no doubt that this condition is part of mainstream veterinary medicine.

Recognising the Process of Brain Ageing

Cognitive dysfunction is the name given to the age related progressive neurodegenerative disorder which is characterised by specific behavioural changes, and physical modifications of the central nervous system, such as the deposition of beta amyloid plaques, production of free radicals and onset of vascular changes, which lead to characteristic neuropathololgy.

Unlike its physical counterparts cognitive dysfunction is characterised by a lack of recognisable clinical symptoms and this can lead to difficulty in identifying cases, especially in the early stages of the condition. Patients suffering from CD primarily exhibit changes in their behaviour and unless owners are asked the right sorts of questions it is easy for the disease to go undetected. Alterations in learning and memory together with changes in social interaction, sleep wake cycles and activity patterns are traditionally associated with the condition and specific changes in the perception of and response to stimuli in the environment, in levels of anxiety and restlessness and in spatial orientation and levels of confusion are also identified.

In order to assist in the clinical evaluation of these cases the behavioural changes associated with this condition are commonly divided into four main groups, namely:

- disorientation
- changes in social and environmental interaction
- changes in sleep/wake cycle
- changes in response to learned associations and commands.

On their own each of these categories could be indicative of a physical disease or could be the result of a purely behavioural condition but when signs are present from each of these categories, and most importantly from the first two, a diagnosis of cognitive dysfunction needs to be considered.

The Human Dimension

A study by Neilson and colleagues in 2001 showed that 28% of dogs in a population aged between 11 and 12 years and 68% of a population aged between 15 and 16 years showed at least one sign which was consistent with a diagnosis of canine cognitive dysfunction and yet only 12 % of pet owners reported these changes to their veterinary practice. One of the reasons for this discrepancy may be that many owners are reluctant to speak to their veterinary practice for fear of being advised to consider euthanasia for their beloved companion. As a result they tolerate changes in their pet's behaviour, seeing them as an inevitable effect of growing old, rather than a symptom of a underlying disease process that can be managed, even if it cannot be cured, and the belief that nothing can be done to change them will drive many owners to make considerable changes in their own lifestyle in order to accommodate their pet's peculiar ways.

Alterations in sleep/wake cycles can take the largest toll on the humans in the household but owners will often put up with broken nights and significant sleep deprivation rather than approach their veterinary practice and run the risk of being persuaded to put their pet to sleep. Breakdown in housetraining can also put an enormous strain on the pet owner relationship but many owners feel guilty about their negative feelings toward their pet and chastise themselves for not being patient with their ageing companion at these difficult times. After all quality of life is a difficult concept and it is important to remember that the criteria by which owners judge the quality of their pet's life will be very different from those that a detached professional might employ. The pet is a valued member of the owner's family and its behaviour is often one of the major indicators of the strength of that relationship as well as one of the most significant sources of strain upon it. However, in the context of old age forgiveness for changes in sleep patterns and breakdowns in housetraining is usually forthcoming from owners who see these as an inconvenience that simply has to be endured.

In contrast changes in social interaction and signs of disorientation are not seen as inconvenient alterations that need to be accommodated, but rather as indicators of the pet's distress. Owners find it upsetting to see their companion looking confused and when their dog fails to respond to previously known commands or their cat startles when they enter the room as if he does not recognise them the effect on the pet-owner relationship can be devastating. Feelings of guilt at the prospect of prolonging life for a pet that is no longer fully aware of its surroundings can be very real but belief that euthanasia is an unacceptable way out can lead to emotional conflict that owners find hard to resolve. Some find it hard to forgive themselves for even considering euthanasia when their pet appears to be in such good physical health, since human society frowns upon the concept of euthanasia and the only way of justifying it in a veterinary context is as a means of stopping physical suffering. Putting to sleep due to changes in behaviour is therefore seen by many owners as a convenience and this only adds to the guilt.

Ageing Felines

Although most of the companion animal literature relating to behavioural changes in old age have been concerned with the dog, some work on the effects of ageing on feline behaviour has also been carried out. In a prevalence study conducted by Landsberg and colleagues in Canada 35% of cats over 11 years of age demonstrated signs of cognitive dysfunction, which were not complicated by any other medical condition, and 30% of cats between 11 and 15 years were affected. In the category of cats older than 15 years the percentage of affected cats increased to 48% and these cats also showed more signs of the condition than their younger counterparts.

The presentation of cognitive dysfunction in a feline context is very similar to the condition in dogs and to related conditions in people, with signs of confusion, disorientation, alteration in social behaviour and breakdowns in house training all being reported. Night time waking, excessive vocalisation and aimless wandering and pacing are also commonly cited in the ageing feline population and changes in sleep/wake cycles appear to be the most commonly reported change in the population of cats older than 15 years.

Asking the Right Questions

Noticing the signs of cognitive dysfunction early on is crucial and there is a much better prognosis in terms of the level of improvement and the extension of good quality life if treatment is instituted in the early stages of this disease. However, it is this early stage that often goes unnoticed and the most effective way of increasing the detection rate for this condition is for elderly pets to have regular veterinary examinations, which include a behavioural a well as a physical health check. Information about the animal's social behaviour, as well its sleep patterns and toileting habits, will lead to early detection of the condition and allow treatment to be instituted as quickly as possible.

Investigation of Canine Cognitive Dysfunction

In order to assist in the early detection of cognitive dysfunction in dogs it can be useful to divide the areas of investigation into the four main categories and then to ask questions which are designed to identify alterations in behaviour. A similar approach will be needed in the investigation of feline cognitive dysfunction but at present far more is known about the subtle signs seen in dogs and information about treatment opportunities is certainly more readily available in the canine field.

DISORIENTATION

Dogs suffering from cognitive dysfunction will show a delay in the recognition of people, places and objects and in some cases there may be no recognition at all. Obviously when dogs fail to recognise their owners this is likely to be noticed quite quickly, but failure to respond to people who call regularly at the house or to those that are met while out on daily exercise may go unnoticed for quite some time. The way in which the dog greets people at home and on walks can help to identify canine cognitive dysfunction in its early stages. Some old dogs suffering from cognitive dysfunction will wander aimlessly around the house and it may be obvious that these individuals are disorientated and confused but in other cases more subtle alterations in the dog's reaction to previously familiar objects, such as household furniture and trees and bushes in the garden, can help to pinpoint the symptoms of cognitive dysfunction. These dogs will often bark at these objects as if they have never been seen before and some may show real fear. Failure to recognise the home driveway on the return from a walk or a tendency to sit at internal doors when asking to go out into the garden can also be symptoms of disorientation and unexplained staring is another potential indicator of this condition.

CHANGES IN SOCIAL AND ENVIRONMENTAL INTERACTION

One of the most obvious signs of a lack of connection between context and behaviour is seen in the interactions that older dogs have with the people and other dogs that they come into contact with. One of the most distressing examples of this is an alteration in the social interaction between the dog and its owners and a decrease in the enthusiasm of greeting behaviour. Together with a decrease in the time spent engaging in play and in affectionate interaction this can signal the onset of age related brain changes. There is often a change in the consistency and speed with which these dogs respond to commands and owners often mistake this change for stubborn behaviour. When considering interactions with other dogs it is not uncommon to see an increase in confrontational reactions in other dogs that appear to be threatened by the bizarre behaviour of the geriatric individual. Dogs suffering from cognitive dysfunction may also become more irritable themselves and owners may be aware of an increase in aggression from their dog together with a general decrease in the desire to interact and play with other dogs on walks.

CHANGES IN SLEEP/WAKE CYCLES
Alterations in sleep/wake cycles are common in cases of canine cognitive dysfunction but unless the dog is disrupting the owner's sleep this aspect may easily go unnoticed. When the dog is whining or barking at night most owners will rapidly respond by taking their pet out into the garden on the assumption that it needs to relieve itself but dogs suffering from cognitive dysfunction will rarely need to eliminate on these occasions. In some cases they may respond to being outside by toileting but once back in the house they will be very slow to settle and will soon recommence barking and whining. Pacing is another common feature in these cases and owners often comment that their dog shows signs of agitation and restlessness at their usual bedtime and paces and vocalises when they make preparations to go to bed. Waking in the middle of the night is another classic symptom of cognitive dysfunction and since these dogs are also disorientated and confused they will often seek out their owners when they wake. If the dog is housed downstairs during the night this

may cause the dog to scratch at doors and this symptom can commonly lead to confusion with separation anxiety. During the day the sleep/wake cycle is also affected and dogs suffering from the behavioural effects of brain ageing will sleep for longer in the day but once again this symptom may easily be overlooked and considered to be a normal change in a dog of advancing years.

LOSS OF PREVIOUSLY LEARNED BEHAVIOURAL RESPONSES

Break down in house training is probably the most documented example of this change in behaviour and can occur in cases of canine cognitive dysfunction for two reasons. Firstly the disorientation associated with the condition can lead to a situation where dogs sit at internal doors when they want to go out into the garden and owners can easily disregard this signal until it is too late. Staring at the hinged side of the door rather than the handle side has also been noted by owners and in these cases they often observe the peculiar behaviour but do not recognise it as a signal for needing access to outdoors. Another factor in the house training problems associated with old age behavioural changes is a failure to maintain associations with suitable latrine substrates and in many cases there is a history of a progressive break down in associations leading to elimination in a wide range of unsuitable locations. These dogs often start by not always toileting on grass as they have done in the past and making mistakes on the patio, on the flowerbeds and eventually on the carpeted floors in the house. In many cases the onset of this problem behaviour is gradual and it is only when the dog is toileting in a number of locations that the owners realise that these are not one-off accidents.

MEDICAL DIFFERENTIALS

In all of the categories of behavioural symptoms for cognitive dysfunction there are a number of medical conditions that need to be considered as potential causes. Deterioration in sensory functions, such as sight and hearing, need to be considered when signs of disorientation, changes in social interaction and changes in sleep/wake cycles are encountered, since an animal that is not fully aware of its surrounding through a lack of sensory input can easily present with signs which resemble those seen in cognitive dysfunction. Cardiovascular and neurological disorders will also need to be ruled out and the influence of pain on activity levels, social interaction and sleep patterns should be investigated. Disorders of the gastrointestinal and urinary tracts will need to be considered in cases where a break down in house training is the major presenting sign and medical conditions which result in increased thirst or increased urination will also need to be included in the investigation. Only once a thorough medical investigation has been carried out by a veterinary surgeon can the behavioural investigation continue and treatment for cognitive dysfunction be considered.

Treatment for Age related Behavioural Changes

Most of the proactive research into the treatment of cognitive dysfunction has been carried out in dogs and at the moment medications and specifically formulated diets are only marketed for dogs suffering from this condition. However, recent work in Canada has suggested that a similar therapeutic approach consisting of a combination of dietary management, medical treatment and behavioural therapy is just as applicable to the feline population.

Medication for Cognitive Dysfunction

Although primarily behavioural in presentation there can be no doubt that cognitive dysfunction is a medical condition and when deciding on medication to treat these cases it is important to consider the changes that are occurring within the body systems including the central nervous system. During the normal ageing process behavioural changes can result from changes in the blood flow to the brain and medication with drugs which enhance the availability of good cerebral circulation can vastly improve this situation and give increased energy and vigour to dogs that are beginning to slow down. In addition there are drugs which lead to a protective effect on the central nervous system and these may be beneficial in cases of canine cognitive dysfunction where internal changes such as a depletion of brain dopamine levels and an increase in the presence of free radicals, leading to cell injury and brain pathology, need to be considered.

In the United Kingdom and the USA there has been a lot of publicity relating to the use of selegiline hydrochloride as a therapeutic approach to cognitive dysfunction and three important actions have been identified in relation to the treatment of these cases. Firstly selegiline enhances brain dopamine concentrations and metabolism, secondly it decreases substances in the brain, which are responsible for neural cell damage, and thirdly it protects nerve cells, decreases cell death, and promotes synthesis of nerve growth factors. It is administered as a once daily dose and in cases of dementia the dog will usually require long term medication. The onset of action of the drug is not immediate and it has been reported that clinical changes have been delayed for up to six weeks after commencing treatment. However others report changes within three weeks. In long standing cases of canine cognitive decline it has been reported that the effects of single drug therapy may not be sustained and that such cases may respond better to combination therapy but in such situations it is important to discuss the dog's behaviour in detail with a veterinary surgeon who can monitor responses to treatment and modify treatment as appropriate.

A Nutritional Approach to Cognitive Dysfunction

It has long been accepted that nutrition can play a vital role in the treatment of organic disease and in cases where behavioural problems in the geriatric pet are attributable to conditions such as renal failure, the use of prescription diets should obviously be considered. However, recently there has been increasing interest in the role of nutrition in the treatment of age related progressive degeneration of the central nervous system and in the clinical consequences of that process. In particular research has looked at the effects of dietary manipulation on canine short-term memory and has assessed this with reference to particular learning and memory tasks. This is particularly significant in relation to treatment of canine cognitive dysfunction since loss of short term memory is known to be one of the first indications of similar conditions in humans. Researchers have found that cognitive performance can be improved with a diet supplemented with a broad spectrum of antioxidants, which are believed to prevent the development of the age-related neuropathology. In addition antioxidants are believed to promote recovery in neurons that are exhibiting signs of neuropathology and therefore commercially available diets and nutritional supplements, which are enhanced with these agents, are believed to offer another option in the treatment of behavioural disorders in the older pet. The role of nutrition in the prevention and treatment of both canine and feline cognitive dysfunction is currently the subject of extensive research and it is likely that further developments will be seen in the near future.

Providing Behavioural Support for a Medical Condition

Although cognitive dysfunction is undoubtedly a medical condition there is a strong behavioural component and behavioural therapy is needed as a support for the pharmacological treatment. One of the consequences of age-related behavioural disorders is the loss of learned responses. As a result dogs may lose the ability to perform simple tasks or respond to previously known commands and cats may show a breakdown in house training or a reluctance to respond to being called in from the garden. Obviously re-establishing these learned responses is particularly relevant in treating cases of canine cognitive dysfunction and teaching these dogs old tricks will involve a large amount of patience and understanding. The use of simple unambiguous commands and clear reward signals is essential. Ideally rewards should be things that the pet particularly values and this will be dependent on the breed and will also vary from individual to individual. For example some dogs will value games, others petting and others food. The use of a clicker, which has been previously associated with reward in a simple introduction process, gives the pet a clear unambiguous signal that will help to reinforce success. Patients will often need to be taught commands from scratch and house training and introduction of very basic obedience commands will be part of the process. The aim of behavioural therapy in these cases is to give structure and predictability to the environment in a way which helps the dog to understand what is expected. Cognitive dysfunction results in a lack of association between action and context and therefore it is important to ensure that all commands are consistent and that the dog is given as many clear signals of success as possible. Visual signals can be very useful, provided the dog is not showing any sensory deficits of course, and marking exit doors can be beneficial for those dogs showing signs of disorientation leading to house soiling problems.

Cognitive dysfunction patients often show difficulty in concentrating and therefore it is important to introduce games that will provide mental stimulation and increase social interaction with their pet. Ideally play and exercise sessions should be of short duration and involve simple tasks, which are repeated frequently and culminate in a positive reward for the pet. For example, several short exercise outings each day will be preferable to one long one since this will stimulate the pet's interest in the environment and provide increased opportunities for interaction between the pet and its owner.

The Effect of Treating Cognitive Dysfunction on the Pet-Owner Relationship

The onset of old age is an inevitable fact of life and for many pets their transition into the ranks of the geriatrics is smooth. However this is not always the case and when signs of disorientation and confusion begin many owners find it hard to recognise their faithful family friend. The changes in social interaction can make brain ageing a distressing condition for the owner and when dealing with these cases it is important to remember the human element. Detecting the symptoms of this condition at the earliest opportunity will enable pets to receive appropriate veterinary care and maximise the benefits of therapy in terms of increased quality and duration of life.

14

Welfare Issues in Companion Animal Behaviour

Donna Brander and Natalie Waran

'Issues of companion animal welfare are often too close to home to be seen by members of the general public.'

Professor Bernard Rollin – Animal Ethicist

Attitudes Towards Animals

The human animal bond is considered to be over 12,000 years old and was probably originally based on a mutually beneficial relationship. Dogs and cats exploited the tribal life of human beings as scavengers for the dross of the nomadic wanderers. The dogs would defend this resource from outsiders and, thus became a helpful alarm against the approach men or beasts. Cats were a natural vermin control. Sometimes dogs would be used as a beast of burden for nomadic peoples, and both dogs and cats could ultimately be used as food. But there is evidence that even from the beginning, these animals were sometimes kept solely as companion animals.

Many tribal people (for example, Native Americans and Aboriginal people) believe animals are individuals. They believe that animals are thinking, feeling beings that are worthy of respect as ancestors and relatives. Some tribes of

Native Americans call them the 'animal people' and consider animals to be part of the circle of life. Interestingly, many of the game laws, closed seasons and limited harvesting of animals that we apply today have roots in tribal thinking.

Various other ways of considering animals have developed through the centuries. Dominionism dictates that animals exist to serve humankind. Stewardship perceives humans as caretakers of other animal life. Evolutionary Perspectivism is problematic for some, as it looks only at the development of life from a human point of view. Critical anthropomorphism recognises the problem of homocentricism and attempts to critically assess behaviour in animals.

Religious beliefs have also played a crucial part in attitudes towards animals. Some religions could almost be considered to be animal liberationist. Buddhism, Jainism, and the yogic branches of Hinduism perceive animals as having rights, and have a basic tenet of kindness towards all animals. The Quakers were the first Christian sect to oppose blood sports and incorporate kindness to animals as an article of faith.

Attitudes towards animals have changed throughout history. Early Christian attitudes considered man to have dominion over all animals. In the middle ages, animals were used for work, food and skins, but by the late middle Ages the line between humans and animals became more blurred, to the extent that animals were often tried for crimes just as if they were human beings.

In the 18th century, reason and intellect set humans apart from animals. The Cartesian view of animals as machines led indirectly to the belief that animals could feel no pain and, therefore, there was no such thing as cruelty to animals. Cruelty to animals infused eighteenth century Britain, and the ill-treatment of animals was a part of social fabric. However, by the 19th century, humans began to define themselves as creatures of feeling and passion, rather than intellect and they therefore became more empathetic towards animals. It becomes clear that what we think about animals largely depends on how we define ourselves.

In the 19th century attitudes towards animals began to change dramatically with the theories of Darwin. Once again, there was a joining of humans to animals. With these philosophies came what is known as the Generosity Paradigm, which suggests that the vulnerable, unprotected, undefended and morally innocent have a moral priority over other competing claims. Along with this thinking came the concept of protection of cruelty towards animals, and organisations such as the American Society for the Prevention to Cruelty to Animals (1866) and the Royal Society for the Prevention of Cruelty to Animals (1824) were formed.

What is Welfare?

There are various definitions of animal welfare but, in general, an animal's welfare relates to its ability to adjust to, or cope successfully with, the prevailing conditions in which it finds itself. As a concept, welfare can be described in terms of the Farm Animal Welfare Council's 'Five Freedoms'. These state that the basic requirements for any animal are food, water, good ventilation and protection from risk of injury or disease. In addition animals should be able to perform their natural behaviour and be free from suffering.

Most pet owners equate welfare with the quality of an

animal's life. The problem is that quality of life can be difficult to accurately determine, since there is no one simple indicator that is universally accepted as a measure of welfare. There are a number of measures that can be used, ranging from measures of the animal's biological functioning, for example its health and reproductive ability, to more subjective assessments of its emotional state, including how free it is to perform natural behaviour. In general most owners tend to presume that their dog is experiencing positive emotional states, such as happiness, when it behaves in a certain way towards them, for example wagging the tail, and showing play behaviour. In this way owners often describe their dog or cat's welfare in anthropomorphic terms. Interestingly, people rarely see themselves as restricting the freedom of their companion animals, for example by placing them in an unnatural setting or restricting their interactions with conspecifics. There is also no attempt of objective assessment of whether these animals are free from unpleasant emotional states.

Sadly it is often the case that the care of pet animals can be misguided, although well intentioned. It is still all too common to consider the care and welfare of animals as being based on a 'common sense' approach. This approach can often lead to pets experiencing levels of care and management that fall short of their basic requirements.

Welfare problems arise when environmental conditions are so extreme that the animal can no longer cope successfully. In dogs and cats such an inability to cope may manifest itself in destructive behaviour, aggression, increased sensitivity to noise, attention seeking behaviour or stereotypical responses. These problem behaviours must be considered as indicators of a welfare problem. What is therefore required is a method of assessment of animal needs that is not homocentric.

Animal Welfare Legislation

In the UK, animals were traditionally viewed as property, and as such were protected because they belonged to the owner. In 1809 in the UK, Martin Erskine presented the first Bill to Parliament that was concerned with the protection of animals against cruel acts. Although this failed to become law, it did encourage others to consider the matter worthy of further debate. It was, however, 1822 before Martin's Act finally became law, making it an offence for any person to beat, abuse or ill-treat any horse, or other beast of burden, and a further four years before dogs became protected under the Act. By 1840, the principle of animal welfare legislation had become accepted, and The Prevention of Cruelty to Animals Act became law in England in 1849. In 1911, the existing anti-cruelty legislation was consolidated and extended by the passing of the Protection of Animals Act. Since then there have been nine amendments to the Act. In essence the Act makes it an offence to cause any 'unnecessary suffering' to any domestic or captive animal.

The problem with this Act is that, as with most legislation, it is concerned with extremes. The statutes tend to focus on animal cruelty, and as such they do not seek to promote higher standards of welfare. Interestingly, in the case of companion animals, UK law is less advanced than the Council of Europe's Convention on pets. The European Convention for the Protection of Pet animals is based on a 'duty of care' philosophy such as is described in the welfare codes for farmed animals. The proposed new UK Animal Welfare legislation based on the European convention has yet to successfully make its way through Parliament.

There are various other pieces of UK legislation that are more specific and aim to control the use, housing and breeding conditions for companion animals. For example the control of dog breeding is covered by The Breeding of Dogs Act 1973, 1991 and the Breeding and Sale of Dogs (Welfare) Act 1999. Under this legislation, anyone who is in the business of breeding and selling dogs requires a license. This is granted only if the animals in the breeding establishment are suitably accommodated, fed, exercised and protected from disease and fire. It is for local authorities, which have extensive powers, to check on the standards of health, welfare and accommodation of the animals, to enforce the requirements of the Act. In addition the 1999 Act provides that bitches may not be mated until they are at least one year old and that they give birth to no more than six litters in a lifetime, and no more than one litter per year.

The Dangerous Dogs Act 1991 is another example of legislation that has welfare implications for pet dogs; however in this case the issues lie with the drafting of the legislation. After a series of well-publicised attacks by dogs on humans, the UK Government hastily responded with an Act that, amongst other things, aimed to control the breeding and movements of certain breeds of dogs. The confusion lay in the interpretation of the wording of the Act, which led to long delays in proving whether a dog was of a certain type, and, as such, whether the owners were subject to prosecution for having custody of it, off-lead or un-muzzled, in a public place. The consequences of this were that many dogs were confiscated from their owners and placed in kennels, with the average length of detention being almost two and half years. At one point the Metropolitan police were holding some 448 dogs under this Act.

Of course animal welfare law is constantly being revisited and revised and new proposed legislation is constantly being put before Parliament. For example a recent controversial Bill relates to the sale or use of collars for animals, which administer electric shocks.

Common Welfare Problems

Typically companion animals are required to fit their owner's lifestyle and additionally, the owner's concept of a social system. Human misunderstanding and high expectation is usually the reason for the development of problem behaviours. Of course companion animal behaviour problems usually impact on the animal's human companions and the behaviour is either addressed or the relationship abandoned. Dog and cat rescue homes provide temporary care and a re-homing service for thousands of pets each year, and one of the reasons for animals being placed in such homes has been shown to be behaviour problems.

PROBLEMS FOR DOGS

Dogs have a different social system to humans, and may have problems of separation from their human 'pack' unless accustomed to the situation. The 'urban dog' is an example of the social pack animal that is left alone without its group members. This causes the animal distress and it may respond by barking, howling or whining in an attempt to deal with the situation. The dog may attempt to escape the confinement or, in extreme distress, urinate, defecate or engage in destruction. Preventing such problem behaviour is the most effective way of avoiding the stress that this imposes upon the dog. During early development it is suggested that the dog be trained to deal with such unnatural situations in order to be comfortable with human lifestyle requirements.

Pet dogs can also exhibit problems with anxiety that can lead to aggression or fearful behaviour. This sort of problem behaviour can be attributed to the lack of human understanding of the different (linear) hierarchical social system required by dogs as compared with the human social system. It can also be caused by a lack of understanding of the importance of exposing puppies to commonly encountered stimuli. Problems arise where the dog is treated as an equal in a relationship, or is expected to share its food bowl, or some other important resource. A dog that is unwittingly allowed to develop a sense of high status may become competitive, which may lead to agonistic behaviour, where-as a low status dog may exhibit signs of increasing fearfulness, such as hiding behaviour.

PROBLEMS FOR CATS

Problems of territory are often the cat's main welfare problem. It has been stated that the single most stressful situation for a cat is another cat. This can lead to problems when a household or neighbourhood has many cats. Even one cat, chosen to keep the resident cat 'company', may cause stress leading to behavioural problems, such as aggression between the cats, aggression towards people and marking in the house by one or both cats. Cats are social and many enjoy having others of their species around. They prefer to choose their own companions and they may not necessarily accept one that has been arbitrarily brought into their life by well-meaning people.

Most cats require many options to live within a territory in peace. A cat's life can be very complex in regard to time spent in various areas. Availability and location of resting areas, movement from one area to another, how food and water is dispensed and where it is dispensed all have to be considered from the perspective of their natural behaviour. For example cats like to snack rather than being fed on a schedule.

An interesting debate that often arises is whether cats require access to outdoors. In the UK, there appears to be a belief that cats need to be able to get outdoors, and, even in urban areas, cats may be shut out of the house during certain periods of the day. Often these time periods relate to the owners lifestyle and not to the cats, and this can lead to welfare problems where the cat is unable to avoid unwanted interactions or close encounters with neighbourhood cats. A cat's lifestyle can have a significant effect on its chances of being killed. A study carried out in Cambridge found that in a sample of 128 cats involved in road traffic accidents, younger male cats that spent more time outdoors during darkness were pre-disposed to such accidents. Neutered females, pedigree cats, and cats kept inside at night were less likely to be killed.

Cats often have specific litter tray requirements such as placement of trays, number of trays, type of trays, and type of litter. Many of the needs of cats are simply not part of the human sensibility and a 'common sense' approach to keeping cats will, at best, only be adequate, and may not address the animals' behavioural and environmental welfare. It is interesting to note that in the not so distant past, it was thought that cats did not feel pain to the same degree as other animals. This judgement was based on the cats' response (or lack of response) to stress and pain. As a prey animal, part of the cat's survival repertoire is to remain still and hidden when stressed or in pain.

Cats commonly hide when they are distressed, and some cats spend a lot of time hiding in the household rather than interacting with family life. Some of these animals may be in extreme distress, or even pain, but may be neglected because this behaviour is perceived as normal for the cat. While it may be normal for cats to be more discreet in their responses than a dog, any cat that spends a lot of time hiding or running away should be considered as behaviourally distressed, or in pain, rather than behaving normally. Lack of knowledge of a species causes inappropriate expectations. People either think cats are just like dogs, and treat them accordingly, or they have a misconception of true cat behaviour. Either way, the welfare of the cat is marginalized.

Feeding

It is interesting to note that the general public does not consider overfeeding animals, whether cats or dogs, a welfare issue. Most veterinarians have had alarmingly fat animals brought into their clinics at one time or another but they are not brought in because the client is concerned about their weight. Often the client simply does not recognise the obese condition as being detrimental to the animal, and will often be very reluctant to decrease the food intake. Fat animals are often seen as being cute and much loved. Thin, gaunt animals will provoke an immediate concern for the animal, even though statistics show many more animals suffer from complications of obesity than die from being too lean. Again, a lack of knowledge as to the needs of an animal, or even how an animal should look when fit, could be the problem.

Problems for Rabbits

More and more households are keeping rabbits as 'house rabbits'. Rabbits are very social animals and keeping them in the house can be a good experience if the needs of the rabbit are considered. Traditionally, rabbits are kept in a small cage in the garden. They are sometimes kept as a single rabbit and sometimes housed together in pairs. Studies done by Dr. Shirley Seaman at the University of Edinburgh suggest that social contact is so important to rabbits that they will work hard in order to achieve even relatively short periods of visual and minimal tactile contact with another rabbit. Management of this contact is crucial.

Whether they are housed in a cage in the garden with other rabbits or kept in the household, rabbits prefer a living situation in which they have some control over the contact they have with other rabbits. Cages should be fitted with a bolthole that they can use if startled, or to avoid visual contact with other rabbits. Rabbits kept in adjacent cages could have some of the shared wall as a wire mesh in order to facilitate eye contact. In the house, rabbits should have a small cat carrier or box opening to the front in order to 'bolt' if frightened or wishing to avoid contact with other animals. Options in social contact make a rabbit feel secure.

Training and Welfare

Some people enjoy competing with animals and most people would suggest that competing with dogs is mutually beneficial to the dog and the people who do the training. This assumes that the trainer uses training techniques based on a thorough knowledge of the way that dogs learn, and has a clear understanding of the dog's behaviour. Even with the correct use of positive reinforcement there can be problems, such as with the timing of the reward, the clarity of the signals and other factors that can cause problems for the dog in making the correct associations. Breakdowns in training can lead to inappropriate use of aversive techniques and even the use of inhumane methods to force the animal to perform the behaviour. The pressure on trainers can be intensified by the need to produce a dog for competition.

All forms of manipulation should be continually reassessed in a training programme. The use of electric shock collars should always be considered unethical but even standard choke collars can be cruel and dangerous. These collars are still in common use but should be considered unnecessary if more appropriate training methods are applied. There is plenty of evidence of the serious injuries that can be inflicted by these collars, not to mention the increased risk of promoting behavioural problems. Their use as a training 'aid' is certainly frowned upon by anyone who wishes to encourage welfare in the handling of dogs.

The use of any of these inhumane training devices or techniques by trainers is mainly based on the fact that they can successfully achieve the desired behaviour. However, we would argue that this is not a valid reason for their use. The effectiveness of training and behavioural therapy should never be based solely on the success of the results. It should always be based on results achieved through the humane, ethical treatment of the animal.

Problems with Breeding

Genetic manipulation of dogs and cats for the entertainment and enjoyment of human beings is another area for serious welfare concern. Dogs in particular have been selectively bred to enhance certain qualities required for hunting or fighting purposes. Some were selectively bred for a small stature, to be able to go down holes in the ground for prey, or fit more comfortably in a person's lap. Much of the more recent genetic manipulation that is popular in the dog is cosmetic. People want the dog to look a certain way, no matter how disastrous this is medically for the dogs involved.

Selective breeding for fashionable traits has serious welfare implications with dogs being bred with extreme deformities in order to fit a breed standard set by purebred dog enthusiast. Of the breeds recognised by the Kennel Club, 148 breeds of purebred dogs collectively exhibit more than 300 genetic deformities and many of these breeds are affected by multiple inheritable problems. It could be argued that the breeders of these dogs deliberately select for characteristics that are known to cause medical problems for the animal.

Some of the more common breeds that are detrimentally effected by cosmetic breeding include the Cavalier King Charles Spaniel, Chow-Chow, the Basset Hound, the Dachshund, the Boxer, the Shar Pei, the Pug, the English Bull Terrier, the St. Bernard and, perhaps the most damaged of all, the English Bulldog. Many of these deformities cause the animal great distress and/or pain throughout a sometimes-shortened life span. Since there is no other purpose for the selective breeding of these animals, this can only be attributed to human vanity.

A short list of the genetic disorders that are a direct result of cosmetic breeding include: dental problems, injury prone ears, breathing disorders, eye problems, shortened tear ducts, cataracts, inwardly turned eyelids, bone and joint failures, heart problems such as leaky valves and heart murmurs, cataracts, slipped kneecaps, spinal problems, hip dysplasia, skin diseases from skin folds, thyroid dysfunction, skull and brain deformities, and the list goes on and on. Professor David Morton, bio-ethicist and veterinarian at the University of Birmingham considers the issue to be one of 'integrity'. An animal has integrity if it is biologically fit in its natural state and capable of surviving in a range of environments.

Many breeds also suffer from behavioural problems that can be a direct consequence of breeding for function. Gundogs were originally bred to be very active dogs capable of, and requiring, a lot of exercise with enhanced retrieving abilities. This can cause problems of possessive aggression, or simply give rise to exercise requirements beyond the means of most people. Terriers were originally bred to be tough and tenacious hunters, not the lap dogs that many are now used as. Border Collies have proved to be exceptional at agility games, as they were originally bred to have high activity levels and heightened vigilance for aiding sheep herding. Now this breed is being bred for enhanced ability in agility competitions, raising questions as to how ethical this practice may be. The natural characteristics of the breed that were so valued in a functional context as sheepdogs, are now being honed through breeding so that they are more compulsive in their desire to chase a flyball.

Cats fare no better. Cosmetic breeding has created cats with no hair (Sphinx) and cats with such long, fine hair that they cannot properly groom themselves (Persian). The Persian also has eye problems inherent with the aesthetically desirable short muzzle and nose. There are cats with foreshortened legs (Munchkin) that cannot jump or climb and very large cats (Maine Coon Cat) that are prone to hip dysplasia. There are also cats with disorders that go along with the genetic deformities such as the tailless cat (Manx) and the potentially fatal gene attached to the Scottish Fold. Most of these animals are truly disabled and they cannot live natural, normal lives.

A new trend is the crossing of wild species of cats with domestic cats, to create an exotic animal for the pet trade. The Bengal cat and the Leopard Spotted Cat have a large percentage of wildcat in their gene pool. They are extremely active animals and they require a lot of attention and enrichment to keep them from being destructive. These cats prefer to be outdoors but, unfortunately, most of these cats are housebound as they are too valuable to allow out of the house. They have been bred mostly for their looks but little regard has been given to the needs of such an animal.

It is also popular to breed rabbits that would most likely not be able to live a natural life on their own. Some have large floppy ears dragging the ground. Others have fur that requires a daily brush from their human companion in order to keep it from matting.

All of these animals have been bred for their looks and little regard has been given to their ability to participate in natural behaviour.

Does anyone care? The proposed new Animal Welfare Bill included the banning of docking of puppies' tails; tighter controls of pet fairs and the exotic pet trade; a new criminal offence of 'likely to cause unnecessary suffering'; increased powers for police to investigate cruelty; longer sentences and more time for cases to be brought to court; and prosecution of breeders who produce genetically defective animals. Lobbying campaigns by breed associations and owners clubs fiercely opposed every one of these proposals. These serious welfare issues, affecting millions of animals, are now in the hands of the Department of Environment, Food and Rural Affairs (DEFRA) and Parliament.

It could be argued that the deliberate breeding of cats and dogs should be eliminated altogether. Thousands of unwanted cats and dogs are abandoned, neglected or euthanised everyday in this country. Yet the breeding continues. Often the animals breed because their carers did want to take responsibility for neutering them. It is perhaps not so surprising that there is therefore a segment of the population who would like to restrict or even ban people from keeping pets. Their reasoning is that they respect these animals too much to allow the practice of irresponsible pet ownership to continue.

Problems of Misconception

There are certainly welfare problems that arise due to owner ignorance as to the behavioural needs and physical requirements of their pet.

It is difficult to assess how much a dog will do without manipulation because, as a social animal, the dog usually wants to maintain the relationship with his owner. This does not necessarily mean that the dog would want to engage in a particular activity if left on his own.

A good example of misconception of an animals needs, is the walk that most dogs are required to take on a daily basis. It is commonly believed that a dog must be walked at least once a day to stay healthy. This is sometimes done with the purpose of calming the dog down through exercise. Sometimes these walks take the form of accompanying a person who is cycling or jogging. If the dog is mature and fit, these forms of exercise are probably beneficial and enjoyed, although it should always be recognised that dogs like a lot of slow, 'sniffy' walks in which to assess the world around them. It should also be pointed out that if the goal is to 'calm the dog down' by cycling/jogging, the dog probably just gets fitter. This has been known to make the dog fitter than the person and in doing so makes the dog more difficult to live with in the house!

Sometimes this perceived requirement to walk the dog is fraught with stress and anxiety for both the animal and the walker. The benefits of a walk can easily outweigh the costs in terms of the stress, if the dog exhibits aggression or is overly fearful. It is often considered that dogs must be walked even if there is a big private garden available to the dog. Little thought is given to engaging the dog's brain by making the private environment more interesting. Perhaps more consideration should be given to games that can be played in the house or garden that provide sufficient exercise as well as mental stimulation. A dog with a behavioural problem involved with being in public, could benefit from a new private regime. It would likely enhance the welfare of the dog, the walker and the general public if 'permission' were given to the individual to stop walking the dog in public until the problem is resolved or under better control. Confirming that it is okay to curtail walks and supplement with other games available can often be of benefit to all concerned.

A further problem is of old and infirm dogs who are required to continue the daily walk, even though it may no longer be enjoyable to them. A dog may continue to attempt this activity because it wishes to be with the human companion, or it may be a lifelong habit. Assumptions about the need and enjoyment of the activity should be assessed, and compromises made to benefit and accommodate the changing needs of the animal.

A potentially serious misconception is that animals have knowledge of wrongdoing which is a welfare issue. Owners often state this because they feel that it is proven by the 'guilty' posture the animal appears to exhibit when confronted. While there is no scientific evidence that animals do not feel guilt, it is most likely the posturing is done because the animal realises the person is behaving in a strange way, even possibly aggressively.

Commonly misunderstood dog behaviour that can lead to an inappropriate owner interpretation includes the submissive grin that is often perceived to be a snarl and the play bow that some have interpreted as aggression. Clearly if we are going to have these animals in our lives, it is imperative that we educate ourselves on their social systems and behaviour.

The Kennel Environment

Dogs or cats may be placed in a kennel or cattery for short periods whilst the owner is on holiday, or in extreme cases, for longer periods if the animal is in quarantine. Alternatively a pet animal might find itself placed in a re-homing centre for various reasons. In all cases the animal is confined in an unfamiliar environment and exposed to fear provoking stimuli such as strangers, noise or isolation. In the UK almost half a million cats are received in shelters each year, and in the US this number has been shown to be as high as 8 million. Changes in diet, social and physical environment all impose a challenge to the animal's physiology and psychology, and there are well documented cases of health problems that result from the animal being unable to meet the environmental challenge.

Cats may display behavioural inhibition, including failure to groom, feed or urinate; they may sleep for longer, hide more or show stereotypic licking or pacing. A more predictable environment where the animal can feel in control is often the best way to improve the welfare of these animals. Understanding the effect of the shelter environment on dogs can help to ensure that facilities for housing and management are improved. This is turn will lead to the development of fewer problem behaviours whilst the dog is awaiting re-homing. Currently 17.5% of the overall shelter dog population are returned after re-homing. This in turn leads to further welfare problems where the dog is separated from a new 'pack' leading to increased anxiety.

Symbiotic or Parasitic?
So is the keeping of animals as companions mutually beneficial or is it exploitation?

This question can lead to great confusion when discussing how we live with these animals. People keep dogs, cats and rabbits for various reasons. Most people keep these animals as 'members of the family', but this can be problematic for both animals and people, as pointed out above. Animals are often treated in a disrespectful manner, for example as surrogate children, or as 'live toys' to be used by the family children. Both of these attitudes can lead to misinterpretation and anthropomorphisation of the animal's behaviour.

It is apparent that animals are often treated as objects rather than subjects. Sometimes described as the 'Disneyfication' of animals, this can lead to problems where animals have human characteristics projected onto them. An extreme form of anthropomorphism, the animal is depicted as speaking human language, and its body is drawn in such a way as to resemble a humans. Of course most people recognise the difference between a cartoon figure and the real animal, but studies and experience have shown that this material can lead to a misinterpretation of the behaviour of animals by children and some adults. This can lead to a limited knowledge of an animals' true behavioural and physical needs, and unrealistic expectations of an animals behavioural responses. When these expectations are not met, people can be disappointed. They may respond with anger or withdrawal from the animal, leading inevitably, to a reduced state of welfare for the animal that may be expected to 'fit-in' regardless of its needs, leading to behaviour problems, or it may even be discarded.

Finally we need to address the question of what motivates a person to keep an animal as a 'pet'. In some cases it has been suggested that the motivation may be the need to dominate and control members of another species. For others it could be that there is a substitution of animal affection for human affection. We need to consider whether we should be able to treat these animals as property to be bought, sold, neutered, fed and confined at our whim. And finally, we need to consider whether keeping animals as pets is a right, or a privilege.

15

Companion Animals
and the Law

Anne McBride

WHAT IS LAW, WHERE DO COMPANION ANIMALS FIT IN?
When professionals consider the relationship between
owners and their companion animals, the subject of
responsible ownership is often raised. This topic concerns
issues of moral responsibility for the individual animal's
welfare and for that of people, and other animals with
which it may have contact. What is less likely to be discussed is how such moral values and actions have been
translated into laws. That is the need for the professional
to be aware of the legal responsibilities of the owner and
ensuring the owner is also clear about these responsibilities. All too many owners fall foul of the law simply
because they were not aware of it, which is no defence in
court. In this case, ignorance is not bliss.

As Broom succinctly put it in his preface to Radford's
2001 book on animal welfare law; the purpose of law 'is
to promote co-operation among individuals and to discourage actions which harm individuals or society at
large'. Laws are mechanisms by which moral actions can
be promoted, and as we change our view of what is moral,
so this is reflected in the law. This has been particularly

evident with respect to the law relating to animal welfare, where changes over the latter half of the 20th century have reflected our increasing knowledge of animal health, behaviour, emotion and cognition, and our need to discourage actions that harm individual animals.

Interestingly, legal advances in animal welfare legislation were initially concerned with commercial aspects of ownership; farm animals, laboratory animals, pet shops and dog breeders. Little changed in the last century with respect to the welfare of animals kept as companions. This may be partly due to the embedded cultural view that 'pet owners love their animals' and thus know what is best and will 'do right by them', and thereby accommodate their welfare needs. However, owners may compromise their animal's welfare simply because they do not know what is needed, have misinterpreted information, or have been misinformed. In some cases, there are aspects of the human part of the relationship which need professional help, in order to improve the welfare of both parties. In the latter situation it is important that animal and human professionals recognise the need to work together in order to safeguard the welfare of both the animal and owner, and thus meet our legal obligations to both species.

The concept and study of welfare is essentially a scientific issue, whose core meaning was summarised by the Farm Animal Welfare Council in 1993 in the Five Freedoms. These are that animals should have

1. *Freedom from thirst, hunger and malnutrition* by ready access to fresh water and a diet to maintain full health and vigour.
2. *Freedom from discomfort* by providing a suitable environment including shelter and a comfortable resting area.

3. *Freedom from pain, injury and disease* by prevention or rapid diagnosis and treatment.

4. *Freedom to express normal behaviour* by providing sufficient space, proper facilities and company of the animal's own kind, as appropriate.

5. *Freedom from fear and distress* by ensuring conditions and treatment which avoid mental suffering.

To be able to meet these freedoms requires knowledge of the animal's ethology (its natural behaviour), social, nutritional and exercise requirements. In addition, an understanding of the appropriate socialisation, social referencing (that is exposure to the environment) and training is needed to help prevent the animal suffering fear and distress in the future.

Were it that merely being an owner was sufficient to provide the knowledge to 'know what is best', then we would see far fewer animals with problem behaviours; physical issues due to inappropriate breeding, such as hip dysplasia; less need for Pet Health advice on obesity and less teeth mal-occlusion in rabbits due to inappropriate diet. These are just a few of the problems that beset companion animals. It is clear that merely the state of being a pet owner, or breeder, is insufficient to provide for an animal's welfare. Simply loving your pet is not enough; having knowledge is essential.

In 1987, The Council of Europe developed the European Convention for the Protection of Pet Animals. This was ratified (made law) in several counties including Belgium, Cyprus, Luxembourg and Portugal, though interestingly not by the UK, the 'nation of animal lovers'. The Convention lays down requirements relating to all aspects of the animal's life, from selection of its parents (its breeding) to its death, including how it is kept and trained. For example, breeders would have a responsibility for the anatomical, physiological and behavioural characteristics that are likely to put at risk the health and welfare of either offspring or female parent. This would mean a responsibility for, amongst other things, adequate and appropriate experiences being provided to young animals during their early development. In addition, the Convention requires that there is the provision of information and education. It went far beyond the then current UK legislation.

However, 2007 was a significant year. It was the year of the introduction of the UK Animal Welfare Act (2006). This is based on the Five Freedoms and is applicable to everyone who owns or cares for an animal. The development of this Act was a substantial task for the legislators as it involved the distillation and consolidation of 20 previous acts, some dating back over 150 years. The Act directly tackles the concept of responsible ownership by requiring the human to have a 'duty of care' and to do all that is reasonable to ensure an animal's welfare needs are met. It also addresses the issue, previously discussed, of owners lacking knowledge or having been misinformed on how to care for their animal properly. Codes of Practice for different species are to be written to provide information on the nutrition, space, social, physical and behavioural needs of the animal concerned. These codes can be revised as welfare science progresses.

The Animal Welfare Act is relevant to the owner or keeper of any animal of any species. There is specific legislation, the Dangerous Wild Animals Act (1976), relating to keeping non-domesticated species such as some of the snakes, spiders, birds, felids and canids. There is also legislation relevant those who keep 'livestock' species such as pigs, sheep and chickens and to those keeping or owning horses. Addressing these in any depth is outside the scope of this chapter, but it is important that professionals and owners/keepers alike keep up to date with their knowledge. For example, it is law that horses and foals of six months of age are microchipped and have a horse passport. There are a few exceptions relating to wild or semi-wild horses living on Exmoor, Dartmoor and the New Forest. Recent changes (2009) to this law means it now applies not just to 'horses' but to equines, including zebra and 'exotic' horse species such as Przewalski. An excellent source of up to date information on the legal obligations of animal owners and keepers is the government's own website at www.direct.gov.uk.

Much of the current UK legislation relating to companion animals is concerned with the behaviour of dogs and how that impinges on society. This is because they are the species that is most commonly kept and taken out into human society. In general the laws address three characteristics of dogs, namely the noise they can make, the faeces they produce and the fact that they can cause injury.

The rest of the chapter will focus on the general legal responsibilities of the pet, and in particular dog, owner, in the course of normal ownership. The text below attempts to translate the law into its underlying principles and general requirements, rather than spell out all the details of each. For more specific detail, readers are advised to consult the suggestions for further reading and sources of information.

The law relating to animals is sometimes ill-defined and there are many anomalies, even contradictions. This reflects the historical, almost organic nature of its development. Animal Law has grown rather than been planned. If one could compare it to the development of a city, it is more like London with its curved streets, blind alleys and almost impenetrable links and cross-streets, rather than the simple, logical block system of New York or Toronto. There is a recognised need for reform, but this is a mammoth task. However, the consolidation of much previous law into the Animal Welfare Act was a significant step in the right direction.

Cruelty

The Animal Welfare Act (2006), and its sister act the Animal Health and Welfare (Scotland) Act (2006), imposes an overarching Duty of Care on animal owners and keepers. This mean they have to take responsibility to do all that is reasonable to prevent the likelihood of an animal suffering. It is based on the Five Freedoms and the owner can be found guilty of causing suffering or being likely to cause suffering by what they do, or indeed what they omit to do. It is a criminal law and, if found guilty, the courts can impose fines of up to £20,000, imprisonment of up to 51 weeks, banning someone from keeping or owning animals either for a period or permanently and order the removal, rehoming or destruction of the animal, and the owner will have a criminal record.

The Act covers many aspects of behaviour which may cause 'unnecessary suffering' to the animal. For example, owners whose animal's physical or mental health is compromised by being overweight, underweight, under or over exercised, inappropriately socialised / socially referenced (and thus fearful), bred with the likelihood of being born with painful conditions or fearful disposition or trained in a manner likely to cause fear may all be causes for prosecution under this law, just as much as it would probably be considered an offence to carry a rabbit by its ears or a dog in a crate that was too small. It is also an offence to use a dog as a draught animal – that is to pull a cart, sled, barrow or anything else on a public highway. So for those who have huskies or other breeds originally used for such a purpose, such activities can only be done on private ground. It is an offence to abandon an animal, either temporarily or permanently, in circumstances which are likely to cause it unnecessary suffering – unless there is a reasonable excuse.

Fighting and baiting of any animal are also covered here, as is giving an animal any harmful drug or substance or operation. Operations on animals can only be carried out by qualified veterinary surgeons. Indeed, one major outcome of the Act is that it is now illegal to dock the tails of puppies, except under very specific circumstances, and docking has been banned outright in Scotland. Likewise, in 2010 Wales banned the use of electronic training aids, commonly known as shock collars, though this is not the case elsewhere in the UK.

Anyone can bring a case under the Animal Welfare Act, be that a single individual or organisation such as the RSPCA. The issue to be decided by the court is whether suffering has been caused or was likely to be caused by the way the animal was kept. The Codes of Practice are being written in order to help both owners/keepers to meet their legal obligations to their animals and the courts decide whether or not they actually did so! Some codes have been completed and at the time of writing (May 2010) those for dogs, cats, rabbits, equines and privately kept non-human primates are available through DEFRA (Department for the Environment, Farming and Rural Affairs) and the Welsh Assembly.

Can You Have a Pet?

It is not always possible to own a pet. Some housing providers, be they private landlords, housing associations or local authority, have restrictions on what type of pet, if any, are allowed to be kept in their accommodation. Indeed, this also applies to leaseholders, and it is worth checking the lease of any property you are considering buying to ensure there is no clause restricting the keeping of animals.

If you are applying for rented accommodation, do ensure that you are aware of the landlord's pet policy and be honest about whether you currently own or intend to own a pet. Ensure that you obtain written permission from your landlord if you want to keep a pet. Not doing so can lead to difficulties and acrimony and can even result in eviction.

Tenants should also be aware that they are usually liable for any damage caused to the property, its fittings and furniture. Be warned it is not just dogs that can cause damage. Cats, birds and rabbits can ruin furniture and wallpaper and rabbits and rodents are not averse to chewing their way through wiring, potentially causing a fire hazard. Snakes and other reptiles are not unknown to be escape artists, turning up in strange places such as the boiler room of the apartment block. A find that is unlikely to be appreciated by the caretaker or landlord.

Many landlords do have pet policies of some sort or another. Some are not clearly written and others may be very restrictive. Research conducted by my colleague, Ian Bryant, and I found that landlords were open to discussion on whether to allow pets and lacked a lot of information on how to proceed to have more 'pet friendly' policies. This led to the collaboration of PATHWAY (Pets and Housing) and PAC (Pets Advisory Committee) in producing guidelines for housing providers which are also of help to other professionals and animal owners (PATHWAY-PAC, 2007).

Nuisance

The Environmental Protection Act 1990 intends to make our daily lives somewhat more pleasant. It relates to all sorts of nuisance, be it noisy neighbours, street litter or nuisance caused by animals. It is an offence to keep any animal in such a place or manner as to cause a human health hazard, noise or other nuisance.

This obviously relates to keeping animals in appropriate accommodation and responsibly disposing of any waste and faeces so as to avoid any health hazard, or nuisance relating to the smell. The Act also has implications for animals that are left alone and cause a noise nuisance, be that a screeching parrot or barking dog. Often owners are unaware that their animal is noisy while they are out, until they receive a visit from the local authority officer. This can cause a lot of distress that could be avoided.

As ever, it is worth taking preventative measures. Owners need to be advised on how to train their animal to be relaxed when it is left alone. Advice on how to do this can be obtained from the veterinary surgeon, from leaflets produced by national rescue organisations or from APBC members. It is important to ensure the animal has had sufficient mental and physical exercise and has appropriate things to occupy them whilst the owner is out. Finally, neighbours can be asked as to whether there are any problems. This is especially useful should the owner's circumstances change, such as when the animal is left due to changes in their work pattern, as the neighbour can let them know if any noise problems develop.

Dogs Out and About

There are several laws that relate to dogs when they are out and about and the owner's responsibilities. First is that the dog should be with someone! A dog which is roaming unaccompanied is basically straying and the local authority has the right to seize it and detain it for seven days or until the owner collects it. After the seven days the animal can be rehomed, sold or humanely destroyed. If the owner is found, they are liable to pay for all expenses incurred because of the dog's detention. Should the owner turn up after the seven days to find the dog as been rehomed, for example, they have no recourse to get the animal back. As far as the law is concerned they have lost the ownership of that animal. This can be extremely distressing.

Thus it is in the owner's interest to ensure the dog cannot escape from the property and, should it do so, that it is easily identified and its owner traceable. This is why it is a legal requirement for all dogs to wear a collar with the owner's name and address on it or on a tag attached to the collar. Any dog that does not have such a visible identification can be seized and treated as a stray. Having your dog tattooed or micro chipped is an additional measure to help ensure that you and your pet can be quickly reunited. However, this does not negate the need for a visible collar and tag. Indeed, many dogs are reunited with their owner by other members of the public, simply because they were wearing a collar and tag. You may have done just that. It is also worth noting that not all authorities have readers that can read all the varieties of microchip available, though this situation is improving. There is logic to the collar and tag rule. As an extra thought, dogs that wear a check chain should also wear a collar and tag. A check chain can slip off and the dog then is falling foul of the law and can be considered a stray, and be permanently lost to the owner.

Dogs that stray onto the road and cause an accident resulting in damage to a person or to property can land their owners with a substantial bill. The owner is liable for all costs associated with that damage, which could be many thousands of pounds. Indeed, owners of dogs, or any other animal, that cause injury or damage may find themselves in court under the Animals Act 1971. This civil law applies to an animal's behaviour wherever it is, at home or out and about, is frequently used as it enables someone to seek compensation. It is the keeper, the person who had charge of the animal at the time the damage was caused, who is liable and this may or may not be the owner of course. So, this has implications for the friend who offers to take the dog for a walk, or the horse or a ride for example! This act also covers worrying of livestock, a term which includes deer, pheasant, partridge and grouse that are not in a wild state, such as park deer. The dog that worries livestock may pay with its life, as livestock owners are legally enti-tled to shoot a dog seen worrying livestock. Some people think worrying means attacking and injuring livestock, but chasing is also considered worrying. Certainly it is very worrying for the animal, such as a sheep, and may cause them to abort lambs or injure themselves in their fearful escape from what is a predator, namely the dog.

Other laws that relate to the predatory aspect of dogs are the Hunting (Wild Mammals) with Dogs Act (2005) and the Countryside and Rights of Way Act (2000). The latter has restrictions on where and when dogs can walk off lead on private land. They must be restrained on a fixed lead of no more than two metres between 1 March and 31 July and at all times when near livestock. This is to protect the livestock and ground nesting birds. The Hunting law is often believed to concern only fox hunting and packs of hounds. Actually, it is relevant to all dog owners and relates to more than two dogs pursuing animals such as deer, fox, or badger on land where the handler(s) of the dogs has no legal right to hunt.

The bottom line then is to ensure that your dog cannot escape and that when you are out it is kept on lead on the roads and lanes, and around livestock and is under control at all times. Remember all dogs are predators at heart and we cannot predict their every move, no matter how well trained. The sight of a cat or the shock of a car backfiring may cause your dog to run into the road and cause an accident. The simple expediency of a few feet of leash is well worth it!

Where Can You Walk Your Dog?

Different areas will have different restrictions on when and where dogs can be walked. These are enacted in local byelaws; details will be available from the Local Council office. Some may prohibit dogs from particular areas, such as children's playgrounds. Such prohibitions may be applicable throughout the year, others may be seasonal, for example those referring to beaches or fishing lakes. Others may state that in certain areas dogs must be kept on lead at all times. The Countryside and Rights of Way Act (2000) extended the land accessible to people to include some private land. Information on where one can walk can be found on www.countrysideaccess.gov.uk. There are, however, some restrictions: a dog must be kept on a lead of no more than two metres when around livestock and between March 1st and July 31st, which is the breeding season for ground-nesting birds such as larks.

Fouling

Dog mess is unpleasant! It is smelly, difficult to get off your shoes and can pose a serious health hazard to people. It is estimated that the 7.4 million dogs in the UK produce some 1000 tonnes every day!!! Whilst it is not an offence to let your dog relieve itself out on a walk, it is an offence to fail to remove the faeces from designated land and / or to dispose of it inappropriately. Designated land is all land open to the air to which the public have access, and which has been labelled as such by the local authority. Exceptions include land alongside a road with a speed limit of greater than 40 miles per hour, agricultural land, woodland, marshland, moorland, heath and some common land. Under the Clean Neighbourhoods and Environment Act (2005), if you do not pick up after your dog you are liable for an on the spot fine or a fixed fine of up to £1,000. A hefty price! My advice is to make sure you always have a supply of plastic bags in your pocket and to dispose of them properly. It is beyond my understanding why people leave bags of dog faeces under hedges or swinging from branches. The bags, like other litter, can be fatal to wildlife and are bad for the very environment which presumably the person came out to enjoy with their dog!! If properly bagged, dog waste can go in any public litter bin. It is illegal to dispose of it in your green waste bin or ordinary rubbish bin. Alternatives are to bury it, have a 'dog loo' in your garden, or a 'dog poo wormery', or simply flush it down the toilet where it will be broken down safely along with the human faeces.

Taking Your Pet Abroad

Under the Pet Travel Scheme, owners can take their pet abroad, for example when going on holiday. They should consider if it is really appropriate to do so, especially if planning to go somewhere hot! Pets may not appreciate the heat of the sun as much as owners do.

If owners do wish to take their pet abroad, they need to plan several months ahead. The animal will need to be micro chipped and vaccinated against rabies. Blood tests will need to be taken to ensure that the vaccine has taken and the animal is protected. Depending on the destination, Import and /or Export certificates may be required. On return to the UK evidence will be needed to show that the animal has been treated for ticks and worms through the production of an appropriate veterinary certificate. There are several parasites that are common to other countries, but are rare or not found in the UK.

The Pet Travel Scheme is frequently updated with respect to the countries included and the species of animal to which it applies. Owners are advised to contact their veterinary surgeon for further advice, or look at the DEFRA website which will provide up-to-date information.

Owners should also be aware that whilst in another country, they would be subject to that country's animal legislation. For example, in some countries it is legal requirement to have certain breeds of dog, or even all dogs, muzzled when in a public place. Owners need to know what laws will apply and to have prepared accordingly, such as having brought an appropriate sized, basket type muzzle and having taught their dog to accept wearing it in a stress free way.

Dangerous Dogs
There are three acts of law that relate to dangerous dogs. These are the Dogs Act 1871, which is a civil law, and the Dangerous Dogs Act 1991 (Amendment Act 1996) and the Control of Dogs (Scotland) 2010 which are criminal laws.

They are intended to ensure that the public is not at risk from dogs. Either can be used to prosecute a case, though the criminal Dangerous Dogs Act is the one that most members of the public have heard of. It is a common misperception that they only relate to certain breeds, such as the Pit Bull Terrier. Whilst there are specific provisions for some breeds in the Dangerous Dogs Act, in fact both acts refer to all dogs, be they toys, giant breeds, crossbreeds or mongrels. The Yorkshire Terrier or Papillion come under this legislation just as much as the German Shepherd, Labrador or Rottweiler.

Under the Dangerous Dogs Act it is a criminal offence to have 'a dog dangerously out of control in a public place', and in Scotland this has been extended to any place, including the dog's own home and garden. However, under all three acts it is an offence to have a dog that is dangerously out of control. Dangerously out of control means there is reasonable cause to believe that a dog might injure, or cause injury to somebody, whether or not it actually does so. So for example, if your dog runs up to a member of the public and bounces around them barking, even jumping up, that person may have reasonable apprehension or cause to believe that they may be injured and so bring a case against you. Amongst other scenarios, a dog being considered out of control could involve the possibility of a person being tripped up, bowled over, scratched or bitten, the dog causing a cyclist to swerve and fall off their bike, being involved in a dog fight or even causing injury by proxy, such as a dog that causes other animals to flee who in turn could cause injury, for example wild ponies near a car park.

If found guilty and no injury has occurred, then the courts may invoke a control order. This could involve the dog being restricted to being on lead, muzzled or banned from being in certain areas. Should an injury occur, no matter how slight, the court may decide to order that the dog is destroyed. It is important to be aware that injury in this case does not just mean the dog has bitten. In either case, injury or no injury, if the case was under the Dangerous Dogs Act, the owner / keeper will have a criminal Conviction, and may even incur a prison sentence. Having a criminal record can have implications for other aspects of a person's life.

This may sound rather harsh and 'anti-dog'. However, we should always consider the wider social context of all legislation. Dogs, like cars, unless controlled can cause damage. Again, prevention is better than cure.

Adequate and appropriate socialisation to people of all kinds; of different ages, mobility and colour, as well as to other dogs, means that a dog is unlikely to cause damage to a person through fear, or be considered to be dangerous by getting into dog fights. Appropriate training of basic manners on how to approach people is also important. Teach dogs to wait for permission to approach people and to sit when it does so. It is also essential that dogs have a strong recall. It is important to practice all of these when the dog is both calm and excited; practice through play and in a variety of situations. It is not uncommon for owners to say their dog is very obedient 'unless s/he is distracted'! Well, if the distraction and consequent lack of obedience was the issue that led everyone to court, the fact that it is obedient at home is unlikely to be considered a defence! Training needs to be started in puppyhood, and continued throughout the animal's life. A well trained, responsive dog will be an ambassador for dogdom, rather than an advert for the less desirable qualities of the canine race.

Breeding of Dogs

In 1999 the Breeding and Sale of Dogs (Welfare) Act was invoked. As its name implies this is intended to improve the welfare of breeding dogs and their puppies. If five or more litters a year are bred, a licence from the local authority is required. The breeding premises will be inspected by a veterinary surgeon before a licence is granted and will be inspected regularly thereafter. The act also has specific restrictions on the breeding and/or sale of puppies. Anyone found to not have a licence will be fined and may face a prison sentence.

However, owners may wish to breed from their dog and do not come under the specifications of the licence. They do need to think carefully however. No matter whether only one litter or several are bred, homes are not always easy to find and even then it is not guaranteed that a puppy, which is very cute when young, will still be wanted when it is older. The number given to rescue societies because they are no longer wanted is testimony to this.

Breeders should also remember they have legal obligations under the Animal Welfare Act regarding the physical and mental welfare of the animals they breed, whatever the species. This includes breeding from healthy animals, and appropriately educating the offspring with respect to people, other animals, the environment and good manners.

Insurance

Unlike with car ownership, there is no legal requirement to have an animal covered by liability insurance. However, it is strongly advised that owners have up-to-date comprehensive liability insurance to cover them against legal costs, legal representation and compensation for damage caused by the animal to other people or property, even the escaped hamster that chews the electric cables. It may be included as part of a household insurance policy and is also available from some pet health insurance companies.

It is also worth considering insuring the health of a pet. There is no National Health Service for pets and veterinary fees can soon mount up. There are many policies available, for a variety of species, and it is worth shopping around to find the one most suitable.

As with all insurance cover, it is important to read the small print and ensure that the policy covers all the relevant issues, be that for health care or for third party liability insurance.

Final Thoughts

This chapter has only briefly covered the legislation that is most commonly enforced with respect to pet ownership. There are other laws that have not been discussed, such as those relating to dogs and game or the keeping of dogs as guard animals. The law is not a static institution, but changes and is refined as our understanding of the animal world progresses. The latter years of the 20th century and start of the 21st have seen many such changes occur and more are under consideration.

Animal owners have responsibilities, some of which are enshrined in law, to their animal and to the wider community. The quality of life of the animals in our care is our responsibility and ours alone. They are totally dependent on us for their physical and mental well-being, and we should consider whether we can meet this responsibility before we acquire any pet. We also need to think to the future, our lives can change in unexpected ways and we are obliged to ensure the animal's welfare continues to be provided for. There are sources of help available to assist owners in both the short and long term. The PAC/PATHWAY document and the SCAS book on Companion Animals and Older Adults (Dono, 2005) are just two places from which to get further information of where help can be obtained.

With respect to the wider community, both human and other animals, we need to ensure that our animals are not likely to cause harm or nuisance. Good breeding and early experience will set the animal on the right path. However, this needs to be followed up with appropriate experiences and training. In particular, dog owners need to be aware of the need to teaching their dog to respond quickly and reliably to basic commands such as come, stay, sit, down, heel and leave. Basic manners are not only courteous, after all not every one likes dogs, but will also help prevent any misconceptions about the dog's intentions! Owners also need to ensure that their dog interacts well with other dogs. If there are any problems, take appropriate measures, which may include keeping it on a lead or muzzled *and* seek help. Remember the earlier help is sought the better; problem behaviour often only worsens if left.

If an owner is not planning to breed from their pet, be it dog, cat or rabbit, then it is advisable to have it neutered. This will prevent unwanted litters. In the case of male dogs, neutering reduces the chances of the dog being involved in fights or roaming and possibly being picked up as a stray.

Responsible ownership is an aspect of our general social responsibilities. The law merely serves to underline this basic concept. We need to remember that, like us, our animal companions are part of a wider community. Taking appropriate action will allow us all to live in harmony and for our animals to be considered as valued members of society.

References and Useful Reading

1. Evolutionary Aspects of Canine Behaviour

REFERENCES
(1) Goodwin, D., Bradshaw, J. W. S. & Wickens, S. M. (1997) Paedomorphosis affects agonistic visual signals of domestic dogs. Animal Behaviour 53: 297–304
(2) Gray, M. M., Nathan B Sutter, Elaine A Ostrander, & Robert K Wayne (2010) The IGF1 small dog haplotype is derived from Middle Eastern grey wolves. BMC Biology 2010, 8:16
(3) Pang, Jun-Shen et al (2009) mt DNA Data indicate a single origin for dogs South of Yangteze River, less than 16300 years ago. Mol. Bol. Evol. 26 (12): 2849–2864

BIBLIOGRAPHY
Clutton-Brock, J Davis, S. J. M. & Valla, F. R (1978) Evidence for the domestication of the dog, 12,000 years ago in Israel. Nature 267: 608–10
Ginsburg, B. E. (1987) The wolf pack as a socio-genetic unit. In Man & Wolf. Frank, H. (Ed). Dr W. Junk Publishers, Dortrecht, Nethelands.
McFarland, D. (1999) Animal Behaviour 3rd Ed. Longman, Harlow, England.
Mech, D, (1997) Ten Years With The Pack. Swan Hill Press, Shrewsbury, England.
Mech, D, (1979) The Wolf: The Ecology and Behaviour of an

Endangered Species. University of Minnesota Press, London.

Miklosi A (2007) Dog Behaviour, Evolution and Cognition. Oxford University Press.

Savolainen, P., Ya-ping Zhang, Jing Luo, J. Lundeberg, T Leitner, (2002) Genetic evidence for an East Asian origin of domestic dogs. Science 298: 1610–13

Vila, C., Maldonado, J. E. & Wayne, R, K. (1999) Phylogenetic relationships, evolution, and diversity of the domestic dog. J. Heredity 90(1): 71–77

Woolpy, J. H (1968) The social organisation of wolves. Nat. Hist. 77 (5): 46–55

2. Foundations of Canine Behaviour

REFERENCES

Appleby D L, Bradshaw J W S and Casey R A.(2002) Relationship between aggressive and avoidance behaviour by dogs and their experience in the first six months of life. Veterinary Record 150: 434–438

Cairns RB & Werboff J, (1967) Behavior Development in the Dog: An Interspecific Analysis. Science, 158:1070–1072

Elliot O and Scott JP, (1961) The development of emotional distress reactions to separation in puppies. J. Genet. Psychol., 99:3–22,

Fox M W (1971) Integrative Development of Brain and Behaviour in the dog. University of Chicago Press: Chicago, USA

Fox M, (1978) The Dog: Its Domestication and Behaviour, New York, Garland, S & P.M. Press, pp 69–89 153–176

Fox M. W., & Stelzner, D., (1967) Effects of early experiences in the dog. Animal Behaviour 15, 377–386

Freedman D G, King J A and Elliot O, (1961) Critical period in the social development of dogs. Science 133: 1016–1017

Pluijmakers J, Appleby D.L., Bradshaw J.W.S. (2003) Sensitive

Periods in The Development of Behavioural Organization in the Dog and the Role Of Emotional Homeostasis.
Proceedings of the 4th International Behavioural Meeting, Caloundra, Australia. University of Sydney, Australia

BIBLIOGRAPHY
Freeman, D. (1991) Barking up the Right Tree. Ringpress Books, UK
Gray J (1971) The Psychology of Fear and Stress. Cambridge University Press: Cambridge, UK
Panksepp J (1998) Affective Neuroscience: The Foundation of Human and Animal Emotions. Oxford University Press: New York, USA
Scott JP and Fuller JL (1965) Genetics and the Social Behaviour of the Dog. University of Chicago Press: Chicago, USA
Vincent JD (1986) Biologie des Passions. Odile Jacob: Paris, France

3. Puppy Classes – A League of Their Own!

REFERENCES
Appleby, D. (1993) The Behaviour of Dogs and Cats, Stanley Paul.
Appleby, D.(1997) Ain't Misbehavin', Broadcast Books.
Campbell, W.E. (1985) Behavior Problems in Dogs, American Veterinary Publications, Inc.
Dunbar, I. (1996) Doctor Dunbar's Good Little Dog Book, James and Kenneth Publishers.
Fogle, B. (1990) The Dog's Mind, Penguin.
Klinghammer, E. Wolf Park, Indiana, USA Personal Communication.
Lorenz, K. (1955) Man Meets Dog, Methuen and Co. Ltd.
McCune, S., McPherson J.A., and Bradshaw, J.W.S. (1995) The Waltham Book of Human Interaction – Benefits and responsibilities of pet ownership, Pergamon.

Peachey, E .(1992) Running Puppy Classes, self published
Peachey, E. (2003) The 'Good Puppy!' booklet, self published
Peachey, E.(1998) The 'Good Dog!' booklet, self published
Peachey, E. (2001) Dog behaviour, Paragon.
Pluijmakers, J., Appleby, D.L and Bradshaw, JWS, (2003) Sensitive periods in the Development of Behavioural Organisation and the Role of Emotional Homeostasis, 4th International Behavioural Meeting, Caloundra, Australia.
Serpell, J. and Jagoe, J.A., (1995), The Domestic Dog, Cambridge University Press.
The Penguin English Dictionary
The Concise Oxford Dictionary

5. *The Importance of Positive Reinforcement*

REFERENCES
APBC, (2002), Annual Review of Cases.
Azrin, N.H., Holz, W.C. & Hake, D.F., (1963) Fixed-ratio punishment, Journal of the Experimental Analysis of Behaviour, 6, pages 141–148
Ben-Michael, J., (1997) Emotional Experiences of Dog Owners Facing Disciplining Situations With Dogs, Proceedings of the First International Conference on Veterinary Behavioural Medicine, United Federation For Animal Welfare
Carlson, N.R. (1994) Physiology of Behaviour, Chapter 11, Emotion and Stress, Allyn and Bacon
Cheyne, J.A., (1969) Punishment and reasoning in the development of self-control, Paper presented at the R.H. Walters Memorial Symposium at the Biennial Meeting of the Society for Research in Child Development.
Dweck , C.S. & Licht, B.G., (1980) Learned helplessness and intellectual achievement, in J Garber & M.E.P. Seligman (Eds.), Human Helplessness: Theory and applications, New York: Academic Press

Gross, R.D., (1992) Psychology – The science of mind and behaviour, 2nd ed, Hodder and Stoughton

Jones-Baade, R.E. (2001) The influence of the owner on the development of aggressive behaviour in dogs, Proceedings of the Third International Congress on Veterinary Behavioural Medicine, United Federation For Animal Welfare

Sears, R.R., Maccoby, E.E. and Levin, H., (1957) Patterns of child rearing, Evanston, IL: Row, Peterson

Solomon, R.L., Turner, L.H. and Lessac, M.S. (1968) Some effects of delay of punishment on resistance to temptation in dogs, Journal of Personality and Social Psychology, 8, pages 233–238

Pavlov, I.P. (1927) Conditioned Reflexes, Oxford University Press

FURTHER READING

Appleby.D, (1997), Ain't Misbehavin', Broadcast Books

Lieberman, D.A., (1993), Learning: Behavior and Cognition, Brooks/Cole Publishing Company

Kershaw, E. (1998) Go Click, Pet Behaviour Centre

6. *Scent Communication in the Dog*

REFERENCES

Ackerman L. (1996) Dog Behavior and Training Veterinary Advice for Owners. T.F.H Publications. USA.

Appleby D. L. (1990) The Good Behaviour Guide. Pet Behaviour Centre.

Appleby D. L. (1997) Ain't Misbehaving. A Good Behaviour Book Guide for Family Dogs. Broadcast books.

O Farrell V. (1992) Manual of Canine Behaviour. 2nd Ed. BSAVA.

Overall K. (1997) Clinical Behavioural Medicine for Small Animals. Moseby Publishing. USA.

Price C. (2001) Understanding the Rescue Dog. Broadcast books UK.

Serpell, J (1995) The Domestic Dog: It's evolution, behaviour and interactions with people. Cambridge University Press.

7. The Relationship Between Emotions and Canine Behaviour Problems

REFERENCES

Appleby D., Pluijmakers J. (2003) Separation anxiety in dogs: the function of homeostasis in its development and treatment. *The Veterinary Clinics of North America*, 33, 321–344

Carlson N.R. (1998) Physiology of Behaviour. Sixth edition, Allyn and Bacon, USA

Frijda N.H. (1988) De Emoties. Een overzicht van onderzoek en theorie. Uitgeverij Bert Bakker, Amsterdam

Jones-Baade R. & McBride A, (1999), A Biological Model For Aggression in dogs, Mondial d' Etholie, Lyon, 28–41

O'Farrell V, (1995) Effects of Owner Personality and Attitudes on Dog Behaviour. In; Serpell, J.A. (Ed) The Domestic Dog; Its Evolution, Behaviour and Interactions With People. Cambridge University Press.

Pluijmakers J., Appleby D. L. and Bradshaw J.W.S. (2003) Sensitive Periods in the Development of Behavioural Organization in the Dog and the Role of Emotional Homeostasis, Proceedings of the 4th International Veterinary Behavioural Meeting, No: 32, 18th–20th August 2003, Caloundra, Australia, 119–126

Rolls E.T. (1990) A Theory of Emotion, and its Application to Understanding the Neural Basis of Emotion. Cognition and Emotion, 4 (3), 161–190

Solomon R.L., Corbit J.D. (1974) An Opponent-process Theory of Emotion: I. Temporal Dynamics of Affect. Psychological Review, 81 (2), 119–145

8. Our Relationship with Dogs

REFERENCES

Malabre, A.L, (1995) Lost Prophets: An Insider's History of the Modern Economists. Harvard Business Press.

Matthew 13.57 King James Bible

Mech, L.D. (2008) Whatever happened to the term Alpha wolf? http://www.wolf.org/wolves/news/iwmag/2008/winter/alpha wolf.pdf

Trut, L. (1999). Early canid domestication: The farm-fox experiment. American Scientist 87, 160–169.

FURTHER READING

Beck, D (2010) Dog Owner or Dog Guardian? http://www.apbc.org.uk/blog/dog_owner_or_dog_guardian

Bradshaw, J.W.S, Blackwell, E.J. & Casey R.A. (2009) Dominance in domestic dogs—useful construct or bad habit? Journal of Veterinary Behavior: Clinical Applications and Research. 4 (3) 135–144.

Coppinger, R & Coppinger, L. (2002) Dogs: a new understanding of canine origin, behaviour and evolution. Chicago, University of Chicago Press.

Dawkins, R. (1989) The Selfish Gene (new edition). Oxford, Oxford University Press.

Eaton, B. (2008) Dominance: Fact or Fiction? Revised edition. Barry Eaton.

Horowitz, A. (2010) Inside of a Dog. London. Simon & Schuster

Mech, L.D. & Boitani, L. (2003) Wolf social ecology. 1–34 in: Mech, L.D. & Boitani, L. (eds) Wolves: Behavior, Ecology and Conservation. Chicago, University of Chicago Press.

Pang JF, Kluetsch C, Zou XJ, Zhang AB, Luo LY, Angleby H, Ardalan A, Ekström C, Sköllermo A, Lundeberg J, Matsumura S, Leitner T, Zhang YP, Savolainen P. (2009)

mtDNA Data Indicates a Single Origin for Dogs South of Yangtze River, less than 16,300 Years Ago, from Numerous Wolves. Molecular Biology and Evolution 26(12):2849–2864

Parker, G.A. (1974) Assessment strategy and the evolution of animal conflicts. Journal of Theoretical Biology. 47. 223–243

Serpell, J. (ed) The domestic dog: its evolution, behaviour and interactions with people. Cambridge. Cambridge University Press.

Shepherd, K. (2002) Development of behaviour, social behaviour and communication in dogs, Chapter 2, in Horwitz, D, Mills, D. & Heath, S. (eds) The BSAVA Manual of Canine and Feline Behavioural Medicine, Gloucester, BSAVA.

Vila, C., Savolainen,P., Maldonado, J.E., Amorim, I.R., Rice, J.E., Honeycutt, R.L., Crandall, K.A., Crandall, K.A., Lundeburg, J., Wayne, R.K. (1997) Multiple and ancient origins of the domestic dog. Science, 276. 1687–1689.

Welfare in Dog Training http://www.dogwelfarecampaign.org/

9. *Feline Behaviour Problems: The influence of natural Behaviour*

FURTHER READING

Bradshaw, John (1992) The Behaviour of the Domestic Cat. CAB International

Turner, D and Bateson, P, eds. (2000) The Domestic Cat – The Biology of its Behaviour Cambridge University Press.

Heath, S. (2000) Why Does My Cat . . . ? Souvenir Press

Lansberg,G. Hunthausen, W, and Ackerman, L. (1997) Handbook of Behaviour Problems of the Dog and Cat. Butterworth Heinemann

Robinson, I (ed) (1995) The Waltham Book of Human Animal Interaction. Pergamon Press

Overall, K. (1997)Clinical Behavioral Medicine for Small Animals. Mosby

10. Behaviour Problems in the Domestic Rabbit

FURTHER READING
Lockley, R.M. (1980) The Private Life of a Rabbit. Book Club Associates
Magnus, Emma (2002) How to Have a Relaxed Rabbit. Pet Behaviour Centre
McBride, Anne (1988) Rabbits and Hares. Whittet Books
McBride, Anne (2000) Why Does My Rabbit Souvenir Press

11. Rage Syndrome: In our dogs' minds?

REFERENCES
Colter, S. B. (1989) Complex partial seizures: Behavioral epilepsy. In Problems in Veterinary Medicine, Vol 1(4): Epilepsy, 619–627, ed. Richard J. Indrieri. Philadelphia, PA: J. B. Lippincott Company.
Delgao-Escueta, A. V., Mattsonb, R. H., Kong, L. et al. (1981) The nature of aggression during epileptic seizures. New England Journal of Medicine 305: 711–716.
Dodman, N. H., Knowles, K. E., Shuster, L., Moon-Fanelli, A. A., Tidwell, A. S. and Keen, C. L. (1996) Behavioral changes associated with suspected complex partial seizures in Bull Terriers. Journal of the American Veterinary Medical Association 208(5): 688–691.
Dodman, N. H., Miczek, K. A., Knowles, K., Thalhammer, J. G. and Shuster, L. (1992) Phenobarbital-responsive episodic dyscontrol (rage) in dogs. Journal of the American Veterinary Medical Association 201(10): 1580–1583.
Fogle, B. (1990) The Dog's Mind. London: Pelham Books.
Hart, B. L. and Hart, L. A. (1985) Canine and Feline Behavioral Therapy. Philadelphia. PA: Lea & Febiger.

Kelley, P. & Lewis, S. (1989) eds. The Brownings, Correspondence. Vol 11. July 1845–January 1846. Winfield, KS. Wedgestone Press.

Kelley, P. & Lewis, S. (1989) eds. The Brownings, Correspondence. Vol 12. January 1846–May 1846. Winfield, KS. Wedgestone Press.

Lloyd Carey, J. (1992) Cocker Spaniels – An Owner's Companion. Wiltshire, UK: The Crowood Press.

Marsh, L. and Krauss, G. L.(2000). Aggression and violence in patients with epilepsy. Epilepsy and Behavior 1: 160–168.

Mugford, R. A. (1984) Aggressive behaviour in the English Cocker Spaniel. Veterinary Annual 24: 310–314.

Neville, P. (1991) Do Dogs Need Shrinks? London: Sidgwick & Jackson.

Podberscek, A. L. and Serpell, J. A. (1996) The English Cocker Spaniel: preliminary findings on aggressive behaviour. Applied Animal Behaviour Science 47: 75–89.

Thomas, W. B. (2000) Idiopathic epilepsy in dogs. Veterinary Clinics of North America: Small Animal Practice 30(1): 183–206.

13. Behavioural Problems in Ageing Pets

FURTHER READING

Cantanzaro T E (1999) Care of the Ageing Pet in Healthcare of the Well Pet, Ed. Jevring C and Cantanzaro T E W B Saunders pp 49–65

Cummings B.J, Su J, Cotman C , White R, Russel M (1993) Beta amyloid accumulation in aged canine brain: A model of early plaques formation in Alzheimer''s disease. Neurobiology of Ageing 14: 547–560

Cummings B.J., Head E., Afagh, A.J., Milgram N.W., Cotman C.W. (1996a) Beta Amyloid accumulation correlates with cognitive dysfunction in the aged canine. Neurobiology of Learning and Memory 66: 11–23

Cummings B.J., Head E., Ruehl,W., Milgram N,W., Cotman C.W., (1996b) The canine as an animal model of human aging and dementia. Neurobiology of Ageing, 17: 259–268

Dodman N H (1998) Geriatric Behavior Problems in Psychopharmacology of Animal Behavior Disorders, Ed. Dodman N H and Shuster L Blackwell Science, Malden, MA pp 279–282

Head E (2001) Brain aging in dogs: Parallels with human brain aging and Alzheimer's disease Vet Ther 2(3) 247–260

Kitani K, Kanai S, Ivy G O, Carrillo M C (1998) Assessing the effects of deprenyl on longevity and antioxidant defences in different animal models Ann NY Acad Sci 854; 291–306

Landsberg G, Hunthausen W and Ackerman L (1997) Handbook of Behaviour Problems of the Dog and Cat Butterworth Heinemann, Oxford. England pp 185–194

Landsberg G M, Moffat K and Head E (2003) Prevalence, clinical signs and treatment options for cognitive dysfunction in cats Proceedings of the fourth International Veterinary Behaviour Meeting Caloundra Australia pp 77–83

Milgram N W, Head E, Weiner E, Thomas E (1994) Cognitive functions and aging in dogs: Acquisition of nonspatial visual tasks Behav Neurosci 108; 57–68

Milgram N W, Estrada J, Ikeda-Douglas C et al (2000) Landmark discrimination learning in aged dogs is improved by treatment with an antioxidant enriched diet. Abstr. Soc. Neurosci 193: 9

Milgam N W, Head E, Cotman C W, Muggenburg B, Zicker S C (2001) Age dependent cognitive dysfunction in canines: Dietary intervention in Proceedings of the 3rd International Congress on Veterinary Behavioural Medicine Ed Overall K L, Mills D., Heath S E, Horwitz D UFAW Wheathampstead England pp 53–57

Nikolov R Dikova M., Nikolova T., Nerobkova L., Garibova T., (1987) Cerebroprotective effect of nicergoline and interference with the anti-hypoxic effect of prostacyclin. Methods and Findings of Experimental Clinical Pharmacology 9: 479–484

Pageat P (2001) Description, clinical and histological validation of the ARCAD score (evaluation of Age Related Cognitive and Affective Disorders) in Proceedings of the 3rd International Congress on Veterinary Behavioural Medicine Ed Overall K L, Mills D S Heath S E, Horwitz D UFAW Wheathampstead England pp 83–88

Penalligon J., (1997) The use of nicergoline in the reversal of behaviour changes due to ageing in dogs: a multi-centre clinical field. In: Proceedings of the 1st International Conference on Veterinary Behavioural Medicine. Ed Mills D.S., Heath S.E, Harrington L.J. UFAW, Potters Bar pp 37–41

Postal J M, Van Gool F V and Consalvi P J (1994) Use of Nicergoline in the ageing dog - in the Proceedings of the 19th WSAVA Congress Durban South Africa p781

Neilson J C, Hart B L, Cliff K D and Ruehl W W (2001) Prevelance of behavioural changes associated with age related cognitive impairement in dogs JAVMA 218 (11): 1787–1791

Ruehl W W, DePaoli A and Bruyetter D (1994) Pretreatment characterization of behavioral and cognitive problems in elderly dogs J Vet Int Med 8 : 178

Ruehl W W and Hart B L (1998) Canine cognitive dysfunction In Psychopharmacology of Animal Behavior Disorders Eds Dodman N H and Shuster L. Blackwell Science Publication Malden USA pp 283–304

14. *Welfare Issues in Companion Animal Behaviour*

REFERENCES

Radford M (2001) Animal Welfare Law in Britain: Regulation and Responsibility. Oxford University Press

Wise Steven M. (2000) Rattling the Cage: Toward Legal Rights for Animals. Perseus Publishing

Bekoff Marc (1998) Encyclopedae of Animal Rights and Animal Welfare. Fitzroy Dearborn Publishers

Serpell James (1996) In the Company of Animals. Cambridge University Press

Waran Natalie (2002) The Welfare of Horses. Kluwer Academic Publishers

15. Companion Animals and the Law

REFERENCES AND FURTHER READING

Dono, J Ed (2005) 'Companion Animals and Older Adults', Society for Companion Animal Studies

PATHWAY – PAC (2007) 'Housing Provider Resource – Practical Guidelines on Pet Management for Housing Providers.' A Pathway and Pets Advisory Committee Publication. Free to download from www.petadvisory.org.uk/pac.php?id=8 or from www.dogstrust.org.uk/az/p/petsandhousing/

McBride, E.A. and Bryant, I. 'Pets, Policies and Tenants – The PATHWAY survey of UK Housing Provision'. 10th International Conference on Human-Animal Interactions, Glasgow, 2004

Radford, M (2001) 'Animal Welfare Law in Britain' Oxford University Press

Sandys-Winsch, G (1993) 'Dog Law Handbook' Shaw and Sons

OTHER SOURCES OF INFORMATION

Citizens Advice Bureau

Local Authority websites

Website of Department of the Environment, Farming and Rural Affairs (DEFRA) defra.gov.uk

www.doglaw.co.uk

Codes of Practice and guidance on the welfare acts are available
at:

www.defra.gov.uk/wildlife-pets

www.wales.gov.uk/topics/environmentcountryside/ahw/animal
welfare/pets/codesofpractice

www.scotland.gov.uk/Topics/farmingrural/Agriculture/animal
welfare/AnimalWelfare/AHWSAGuidance/guidancepdf

www.opsi.gov.uk/legislation This site provides the legal acts.

www.endangereddogs.com is a useful site for explanation of legislation relevant to dog owners.

www.newc.co.uk The National Equine Welfare Council has a lot
of information for horse owners

ASSOCIATION OF PET
BEHAVIOUR COUNSELLORS

About the APBC

The treatment of behaviour problems in cats, dogs, horses and other companion animals is a highly specialised field that requires qualified professional help. The APBC represents a network of behaviour counsellors that have achieved the highest proven academic and practical standards available in the field of companion animal behavioural therapy. APBC members abide by a strict code of conduct and continually develop their professional knowledge in the light of new research so that clients and the veterinary surgeons that refer them can be assured they receive the latest expert advice at a reasonable cost. The APBC is also proud of its role as an education provider and continues to promote the practice of pet behaviour therapy to improve the welfare of all companion animals.

The aims of the APBC are:

1. To maintain and ensure the highest professional standards of the practice of companion animal behaviour therapy so that clients can be assured of effective, humane and appropriate advice at a reasonable cost.
2. To establish and promote the practice of companion animal behaviour therapy as a recognised profession.

3. To act as a forum for the exchange of ideas and information about companion animal behaviour and offer advice and information to all interested parties in order to foster a greater understanding of companion animal behaviour, consequently improving their welfare.
4. To provide support for those currently studying companion animal behaviour.
5. To be recognised as the official body that represents practising companion animal behaviour counsellors.

The APBC promotes Full Members as those who have achieved the appropriate standards in order to provide assurance to the veterinary profession wishing to refer cases and to the pet owning public. It offers a network of specialist counsellors throughout the UK and internationally.

In light of the growing number of academic courses in the field of companion animal behaviour and other biological sciences, coupled with the ever-increasing establishment of pet behaviour counselling as a profession, the Association offers four levels of membership, Student, Provisional, Academic and Full. Student and Provisional membership levels complement progression towards certification as a clinical animal behaviourist (CCAB), so that the APBC can support individuals undertaking this career path.

PO BOX 46
Worcester
WR8 9YS
UK
Telephone +44 (0) 1386 751151
Fax +44 (0) 1386 750743
info@apbc.org.uk
www.apbc.org.uk

Useful Contacts

Association for the Study of Animal Behaviour (ASAB)
ASAB was founded in 1936 to promote the study of animal behaviour, and membership is open to all who share this interest. There are now approximately 2000 members, mostly drawn from Britain and Europe. Many members are professional biologists who work in universities, research institutes or schools.

Certification of Clinical Animal Behaviourists (CCAB)
ASAB launched a certification scheme for practicing pet behaviour counsellors in 2003, believing that accreditation by a body independent of any organisation representing the interests of practitioners is the means by which it can be demonstrated to the public and to other professions, such as veterinarians, that certain individuals meet the minimum standards of education, experience and ethics required of a professional clinical animal behaviourist. The APBC was represented on the working party of relevant organizations that worked together to create this scheme.
   http://asab.nottingham.ac.uk/

The Companion Animal Behaviour Therapy Study Group (CABTSG)
Affiliated to the British Small Animal Veterinary Association, CABTSG is open to veterinary surgeons, veterinary nurses and others practising in the field of pet behaviour. It exists to facilitate the exchange of information, experiences and ideas between its members and to this end holds an annual study day and publishes a newsletter for its members.

www.cabtsg.org.uk

European Society of Veterinary Clinical Ethology (ESVCE)
The ESCVE is a non-profit making organisation which aims to promote and support scientific progress in veterinary behaviour medicine and comparative clinical ethology.

www.esvce.org

Contributors

David Appleby MSc, CCAB
Prior to becoming a pet behaviour counsellor David worked with Royal Air Force Police Dogs and was a Guide Dog Mobility Instructor. David has been in practice since 1986. Based at the Pet Behaviour Centre from 1989 he ran regional clinics at veterinary referral centres and for over twenty years and was the visiting clinical animal behaviourist at Queen's Veterinary School, Cambridge University. David is a founder member of the Association of Pet Behaviour Counsellors and a Certified Clinical Animal Behaviourist under the Association for the Study of Animal Behaviour accreditation scheme. A frequent lecturer in the UK and other European countries, David lectures on the postgraduate course for Companion Animal Behaviour at Southampton University. In 2009 David emigrated to Finland where he and Tuulia Burgherr run the behaviour clinic and lecture on applied animal behaviour to undergraduates and residents at the faculty of Veterinary Medicine, Helsinki University. They also have a clinic at AB Klinikka at the EHYT centre, Salo.

Jan Hoole PhD, Dip CABC
Jan is based in Shropshire, where for 10 years she ran Ashley Pet Behaviour Centre, part of the Pet Behaviour

Centre® group of practices. She also taught part time at the University of Central England and gave guest lectures at Harper Adams University and the Merseyside Police Dog Training Unit. She now teaches at Keele University in the School of Life Sciences and the School of Medicine. Although Jan teaches genetics, physiology, and human evolution, among other subjects, animal behaviour remains her real area of interest. Jan shares her life with a husband, two dogs, sixteen chickens and a horse.

Jolanda Pluijmakers PhD, DipCABC
Jolanda holds a post-graduate diploma in Companion Animal Behaviour Counselling of the University of Southampton. In 2007 she graduated with a PhD at the faculty of Veterinary Medicine at Bristol University. Her reseach topic was: 'Developmental and other predisposing factors contributing to behavioural disorders related to fear and anxiety in the domestic dog'. She runs referral clinics at her practice, the Davalon Animal Behaviour Clinic in the Netherlands, and lectures at the Postgraduate Course in Applied Animal Behaviour at the Kaho Sint-Lieven in Belgium.

Erica Peachey BSc(Hons)
Erica Peachey graduated from Hull University with a BSc.(Hons) in Psychology, specialising in Applied Animal Ethology. She then went to the Royal (Dick) School of Veterinary Studies as a Research Associate, to investigate the Behaviour Effects of Spaying Bitches. As prevention is better than cure Erica runs classes for pet dogs, aiming to teach owners and dogs to have a better understanding of each other, to avoid problems and to teach response to basic commands. As well as pet behaviour counselling she also

offers individual training sessions and rehabilitation training. Her special interests include puppy education and
children and pets. Erica frequently acts as an expert witness
in court cases relating to dogs and is a regular contributor
to the written and spoken media.

Inga MacKellar MSc, CCAB

Inga was involved in fostering cats and dogs for many
years for various charities, through which she gained an
interest in animal behaviour. She worked in marketing and
advertising but decided to switch her career focus, as a
mature student, to companion animal behaviour counselling. She completed the postgraduate behaviour course
at Southampton University and went on to achieve an MSc
and has been practicing in the South East of England since
1998. Inga is also a Certified Clinical Animal Behaviourist,
as accredited by ASAB.

She was the Press Officer for the APBC for eight years
and often writes for national press as well as being
involved in radio and TV. She is the behaviourist for Cat
World Magazine and the Pet Plan insurance magazine.

Inga lives in rural Sussex with her five dogs and three
cats and a very patient and supportive husband.

Rosie Barclay BSc (Hons), MPhil, CCAB

Rosie Barclay is a Certificated Clinical Animal Behaviourist
(CCAB) and a committee member of The Association of Pet
Behaviour Counsellors. She holds a BSc(Hons) and an
MPhil in Animal Behaviour and Welfare and is presently
consulting as a companion animal behaviourist in Jersey
(Channel Islands). Rosie regularly appeared on Channel
Four's 'Pet Rescue' and has been featured in numerous
newspaper and magazine articles. She has also published a

book on dog behaviour designed for pet owners called 'Good Dog? Bad Dog?' In her spare time Rosie acts and writes for drama and film as well as working at Durrell Conservation Trust where she is currently welfare training Red River Hogs and Sulawesi Crested Black Macaques. She has two children, a gypsy horse, an Airedale dog, two Maine Coon cats, three chinchillas, six rare breed chickens, loads of fish and a very patient husband.

Charlotte Nevison PhD, BSc(Hons)
Charlotte (Charlie) holds a BSc(Hons) in Zoology and a PhD in the Behavioural Assessment of Animal Welfare from the University of Nottingham. She runs the BSc (Hons) in Animal Behaviour and Animal Welfare at Anglia Ruskin University, Cambridge. Her research focuses on questions relating to how animals percieve the world and how this influences their welfare and management in captivite and domestic situations. Her research subjects have included laboratory animals, dogs, zoo species and horses. Charlie is a member of the ZSl Animal Welfare Committee, is a former Secretary of the Ethics Committee for the Association for the Study of Animal Behaviour and a past Secretary for the International Society for Applied Ethology.

David Ryan PG Dip (CABC), CCAB
David Ryan, current chair of the Association of Pet Behaviour Counsellors, is a Home Office accredited police dog and explosives search dog instructor with twenty-six years of experience in training police dogs and handlers before retiring from the police service in 2007. A holder of Southampton University's postgraduate diploma in Companion Animal Behaviour Counselling, with distinction,

since 2002 and accredited as a Clinical Animal Behaviourist by the Association for the Study of Animal Behaviour in 2008, David now works full time accepting behavioural referrals from veterinary surgeons and provides an expert witness legal consultancy in both civil and criminal litigation involving dogs.

David's blend of practical experience and theoretical knowledge of canine behaviour fuel his particular interest in the development of inherited motor patterns and the ways in which they contribute to pets' emotional balance.

Sarah Heath BVSc, Dip.ECVBM-CA, CCAB, MRCVS
Sarah qualified from Bristol University and spent four years in mixed general practice before setting up a behaviour medicine referral practice in 1992. She is an Honorary Lecturer in small animal behavioural medicine at Liverpool University Veterinary school and a Certified Clinical Animal Behaviourist under the ASAB accreditation scheme. She was an external tutor on the Postgraduate Diploma/MSc course in Companion Animal Behaviour Counselling at Southampton University from the inception of the course until 2008. Sarah is Immediate Past President of the European Society for Veterinary Clinical Ethology and in 2002 became a Founding Diplomate of the European College of Veterinary Behavioural Medicine – Companion Animals and served as President of that College from 2002 to 2008. Sarah has a particular interest in the interplay between behaviour and physical illness in dogs and cats. Her interests in the field of feline medicine include the role of behavioural factors in feline obesity, lower urinary tract disease, cognitive dysfunction and infectious disease. She -

lectures extensively at home and abroad on behavioural topics.

E. Anne McBride PhD, BSc (Hons)
Anne holds a BSc. (Hons) degree in Psychology awarded by University College London in 1978. She was awarded her Doctorate in animal behaviour (Aspects of Social and Parental Behaviour in the European Rabbit) from the same institution in 1986. In 1992 she obtained a Certificate in Conservation and Ecology from Birkbeck College, London. She is a Fellow of the Royal Society of Arts.

Anne has been a practising animal behaviour therapist since 1987 and was the senior behaviour counsellor at the Animal Behaviour Clinic at the University of Southampton which ran from 1998-2009. She ran puppy socialisation and pet dog training classes in London from 1989 to 1998. She is a member of the Association of Pet Behaviour Counsellors and an honorary member of the UK Registry of Canine Behaviourists. In addition to seeing behaviour cases, she writes about and lectures on, nationally and internationally, various aspects of animal behaviour and the human-animal bond, including the law where it is pertinent to pet owners. In 2003 she co-founded, with Emma Magnus and Georgie Hearne, the Rabbit Behaviour Advisory Group. She is the Director of the Applied Animal Behaviour Unit and has been the Academic Director of the Postgraduate / MSc in Companion Animal Behaviour Counselling at the University of Southampton since the course started in 1994. This was the first academically recognised qualification in this subject in Britain. She is a member of the Companion Animal Welfare Council, an independent advisory body.

Emma Magnus BSc (Hons) MSc CCAB
Emma has been practising as a pet behaviour counsellor since 1995. She has run monthly referral clinics at veterinary practices within East Anglia and North London since 1996 seeing owners of dogs, cats, rabbits and the occasional parrot.

Emma is also Features Editor of the small livestock journal Fur & Feather. She has a regular 'Rabbits on the Couch' column within Rabbiting On, the journal of the Rabbit Welfare Fund. In 2003 she helped form the Rabbit Behaviour Advisory Group (www.rabbitbehaviour.co.uk) with Georgie Hearne and Dr Anne McBride.

Georgie Hearne MSc
Georgie practiced as a certified clinical animal behaviourist in London with a special interest in rabbit behaviour and human-animal interaction. In 2007 she and her family moved to Mudgee in rural Australia where they now run an organic farm with free range pigs, poultry, cattle and sheep. They also have plenty of wild rabbits! She teaches puppy classes at the local vet clinics and now concentrates on the behaviour and welfare of the animals on the farm. She remains an overseas member of Association of Pet Behaviour Counsellors and is also a member of International Society for Equitation Science (ISES).

Anthony L. Podberscek PhD
Anthony received his veterinary degree and Ph.D. (in animal behavior and human–animal interactions) from the University of Queensland, Australia. Since 1992, he has been a post-doctoral research associate in the Department of Veterinary Medicine at the University of Cambridge. Since 1997, he has also been the editor-in-chief of

Anthrozoös, a multidisciplinary journal of the interactions of people and animals. His research interests include companion animal behaviour, the treatment of animal behavioural problems, attitudes to animals and animal welfare, and cultural aspects of human–animal interactions. He is a Board member of the International Society for Anthrozoology and is on the editorial board of the Journal of Applied Animal Welfare Science.

Julie Bedford BSc(Hons), PGCE, PG. Dip (CABC)
Julie has grown up and worked with dogs and horses her entire life. She has run a private behaviour practice for 22 years ongoing and has worked for The Blue Cross for some 17 years, initially as a Centre Manager, then animal behaviourist and since 2002 as the Head of Animal Behaviour Services. Her spare time is spent hill walking with her constant companions Harry and Sally the Border Collies.

Donna Brander BSc(Hons), CCAB
Originally from Texas, Donna has run behavioural practices in the USA, Europe and Hong Kong and has been settled in Edinburgh since 1991. In 1995 Donna was appointed as an Honorary Fellow of the Royal (Dick) Veterinary School. Donna established a consultancy at the Veterinary school which she led until retirement in 2005. Her particular interest has been animal behaviour and welfare and she was a frequent lecturer on that subject. Donna achieved a CCAB (Certification Clinical Animal Behaviour) from ASAB in 2004. Since retiring from her consultancy, Donna has pursued her interesting in horse welfare having helped establish a company to design and research a new welfare friendly concept in saddles. Donna also works

closely as welfare officer with the charity 'Cancer and Bio-Detection Dogs'.

Natalie Waran BSc(Hons), PhD
Natalie is a leading authority on horse behaviour and dealing with problem behaviour. She has compiled and edited an academic book on horse welfare (see The welfare of horses (2002) N. Waran (ed), Kluwer Academic Publishers) as well as contributing chapters to various books (eg. Social Behaviour of horses in Social behaviour in farm animals (2001), Keeling L and Gonyou H (eds), CABI; and Training and behavioural rehabilitation in the horse in: The behaviour of the domestic horse, Mills D and McDonnell S (eds), CUP (2004)). She has published both in academic journals and equestrian magazines. Before emigrating to New Zealand in 2005 to become Professor of Animal Welfare at Unitec, she lead an animal behaviour and welfare research group in the Division of Veterinary Clinical Medicine at the University of Edinburgh where she ran the equine behaviour clinic.

Index